THE RAILWAY BUILDERS

Other books by the author

THE CANAL BUILDERS
REMAINS OF A REVOLUTION
CANAL (with Derek Pratt)
THE MINERS
THE PAST AT WORK
THE RAINHILL STORY
THE PAST AFLOAT
THE NATIONAL TRUST GUIDE TO OUR INDUSTRIAL PAST
THE RISE AND FALL OF KING COTTON
WALKING THE LINE
BRITAIN'S LIGHT RAILWAYS (with John Morgan)
STEAMING THROUGH BRITAIN
THE GREAT DAYS OF THE CANALS

THE RAILWAY BUILDERS

Anthony Burton

JOHN MURRAY

First published in 1992
by John Murray (Publishers) Ltd.,
50 Albemarle Street, London W1X 4BD

A catalogue record for this book is available from the British Library

ISBN 0-7195-5084 X

Typeset in 11½/13½ pt Times Roman
by Colset Pte Ltd, Singapore
Printed and bound in Great Britain
by Cambridge University Press, Cambridge

Contents

Acknowledgements

Many institutions and individuals were helpful in the preparation of this book. I owe a particular debt of gratitude to the following where a good deal of research was done: the National Railway Museum at York, the Public Record Office, Kew, and the University of Bristol.

The book is not all my own work: my wife, Pip, helped with the archival research and in the preparation of the manuscript. Her enthusiasm for the project is not entirely unrelated to the family link back to the Brunels.

Illustrations

(*between pages 116 and 117*)

The author and publisher would like to thank the following for permission to reproduce illustrations: Plates 3, 4 and 22, Elton Collection, Ironbridge Gorge Museum Trust; Plate 5, Scottish Record Office and British Railways Board; Plate 6, Borough of Darlington Art Gallery, photograph by Gordon Coates, Darlington; Plates 7, 8, 9, 10, 11, 12, 13, 18, 21 and 24, National Railway Museum, York; Plates 14, 15, 16, 17, 19, 20 and 23, Leicestershire Record Office. The illustrations which appear at the beginning of each chapter are drawings of the building of the Newcastle and Carlisle Railway in 1835 by J.W. Carmichael and are reproduced courtesy of Tullie House, City Museum and Art Gallery, Carlisle.

Foreword

This book had its origins in a conversation that took place some time in 1970. I was discussing what I might be writing next with my agent Murray Pollinger, and was not making much progress, when Murray asked if there was any recent book which I wished I had written myself. I replied, without hesitation, 'Terry Coleman's *The Railway Navvies*'. That set a train of thought in progress which took me back in time from the nineteenth to the eighteenth century and from railway navvies to the canal navvies. There is, alas, a dearth of material on the life of the canal navvies, so I expanded the subject to take in all aspects of canal construction from the promoters to the men with pick and shovel. The result was *The Canal Builders*. After twenty years of working in the field of transport and industrial history, I felt I was finally in a position to attempt the same treatment for the railways, and here it is. The problem has not been a shortage of material but rather an excess of it. From the start I set myself certain limitations – and other limitations were imposed by the material itself.

First, this is a book about civil engineering – the building of the system with its cuttings, embankments, tunnels and viaducts. It is all about the efforts involved in creating a railway system in an inconveniently uneven and lumpy landscape: the locomotives and rolling stock that were to use the finished track have been left on the periphery. Other essential parts of the railway have also been excluded, signalling being the most glaring omission. But it seemed to me that this was too large a topic to introduce here and that if it was to be done properly then it would take up a disproportionate amount of space. I saw this book from the beginning as an essentially human story about the men who built the lines, and I wanted to be sure that that was where the focus of attention was fixed. There are a number of excellent books giving details of the history of signalling, so rather than oversimplify a complex subject I have omitted it altogether.

The next important decision was to think about the time-scale to be covered. Eventually, I decided to give a brief note only on the tramway system that was the forerunner of the modern railways. Tramways are like the later railways only in so far as both involve vehicles running on railed tracks. The tramways were generally private industrial concerns, built with local labour and seldom requiring much in the way of civil engineering work. Because cable haulage by stationary engines was widely employed, hills and valleys could be overcome without extensive earthworks, bridges or tunnels. The Causey arch, built for the Tanfield Tramway in the 1720s, is an impressive exception with its single arch of 103-ft. span. But it is remarkable simply because it was an exception to the general rule: few other tramways required anything on that scale. This is not to underestimate the importance of the tramway system, for there were many innovations introduced there which were to prove invaluable as models for a later generation of railway engineers – structures such as the iron bridges which were such a feature of tramways in South Wales. They remain, nevertheless, works on a minor scale. They are the overture to a far greater work, introducing themes that would be fully developed later.

The start, then, comes with the marriage of the steam locomotive to the railed track on a public railway, but where should it end? In a sense the story has no end. The 1980s saw the start of a project that was first conceived in 1835 and promoted with immense enthusiasm by one of the great entrepreneurs of the late nineteenth century, Edward Watkin – the Channel tunnel. Watkin called on the man who was to be chief engineer of the Severn tunnel, John Hawkshaw, to test the feasibility of the notion – but there was to be no real test. It was not engineering, cost or the will to build that killed the scheme, but politics. Politicians might have been won over but there was one implacable opponent who remained immovable, Queen Victoria. The scheme had to wait for the new age of the European Community to reach fruition, and even now such vital factors as the routes from the tunnel are being decided largely on the basis of political expediency. Victorian engineers would have easily understood the process by which routes have to be bent to meet the demands of powerful pressure groups – whether the landed gentry of their day or the constituents of marginal parliamentary seats in our own.

Fascinating as much of the modern story may be, it seems to me that there was a period, roughly centred on the Railway Mania years of the 1840s, when the basis of the modern rail network was laid down, and which is remarkably coherent. Private companies obtained Acts of Parliament, raised money from the general public and set to work building railways using the maximum of human muscle power and the minimum of mechanical power. It was an age that had something of the primitive heroism of the building of the pyramids, the same reliance on vast human resources. The rate at which the system expanded was astonishing. The newspaper *Railway Intelligence* at the end of 1853 gave some telling statistics about what had been achieved in the previous decade. Up to 1843 about 2,000 miles of railway had been opened – in the next ten years over 5,600 miles were added to that. To put these figures in perspective, in the one peak year of 1848 a total of 1,182 miles were opened – compared with under 1,500 miles of motorway in the first *thirty* years of construction.

The choice of lines from which examples have been taken may seem arbitrary – and in many ways it is – for it was decided not from personal choice, nor on the basis of the importance of a route in the overall system. The choice was determined largely on the basis of the availability of appropriate material. Some companies' early records covering the construction period no longer exist; others have vast ledgers full of minutes which, alas, offer simply a bare recital of facts which in the end amounts to little more than 'work proceeding steadily' with not a hint of what that work might entail. Of the seventy or so railway companies' papers inspected, only a fraction were used. The rest were neglected either because they duplicated material more clearly set out in other documents, because little of interest emerged, or, in one solitary case, because when the relevant papers appeared from the bowels of the Public Record Office, the ink had entirely faded from view and what was left was a large handsomely bound volume of wholly blank pages! But the documents which have not been quoted, and which do not appear in the notes, had their value. It was only by reading many more than I finally used that I was able to feel confident that material that was quoted was reasonably typical of the whole. The portrait of railway building that follows is, I believe, a true likeness.

CHAPTER ONE

Beginnings

'When I came to Raby,' said his Grace, 'I remarked to Mr. Scathy, Oh! here now is a fine track of country, and no Railways to interfere; the Northern part of this country is dreadfully cut up with Railways, but there are none here, nor I hope ever will be, no one will ever think of making one here. I shall have nothing to interfere with my comfort.'

Duke of Cleveland[1]

The Duke, like many a landowner, was to find that his comfort was not so secure as he had thought. In 1842, a deputation arrived on his doorstep with plans for a railway from Barnard Castle to Darlington that would pass through his Raby estate. He was not amused. Around the country other deputations were calling on other landowners, some to be greeted with enthusiasm, others with dismay. But however much the gentry might deplore these intrusions into their lands, they were to find that the movement for building more and more lines was now unstoppable. The Railway Age had well and truly arrived. But when did it all begin? That question has no simple answer.

The antiquarian might turn his learned gaze on routes in Babylonia or rutted tracks to ancient Rome, but it seems unlikely that British engineers spent very much time in such esoteric studies. Yet there are connections: the artificial ruts in which the wagon wheels of Greece ran some 2,000 years ago were set between 4 ft. 6 in. and 5 ft. apart, not so very different from the standard gauge of Britain's railways. There was no deliberate attempt to copy ancient forms, but somehow, it seems, practices have a habit of proving to be self-perpetuating. History may seem to move in leaps and bounds, but stand far enough away and a pattern will often emerge. So it does with railways, as ruts cut in stone gave way to rails of wood and then to rails of iron, until a point is arrived at somewhere in the eighteenth century when not only is the overall pattern clear, but the small steps which moved the whole process forward begin to seem plain.

However well the early roads and rails may have been made, they were no match for the efficiency of waterways, whether rivers or canals. So engineers looked for the best and cheapest way to move goods down to the water, and nowhere was this search more intense than among the collieries of north-east England. London was the great customer for the coal of Durham and Northumberland. Barges could take the coal down the Tyne or Wear to the waiting coaster, and the primitive railways could take care of the first stage of the journey from pithead to river bank. The engineers who built these early plateways and tramways had a comparatively easy task. The route from mine to river was largely downhill, and the railways had a one-way traffic. Faced by a steep hill, the problem was easily solved: the weight of loaded wagons running down the slope could be used to raise the returning empties back up it. In between, horses could haul the trucks along the level sections. So the system evolved: the rails were mounted on parallel rows of stone blocks, which left a clear space down the middle for the horses' hooves. And a system that worked for the rivers of the north could be extended to one serving the canals that began to cover the country in the second half of the eighteenth century. Even if traffic was no longer all in the one direction, wagons could still be hauled up and down slopes by means of the other great invention of the age, the steam engine. In its sturdy engine-house at the top of the hill the giant engine would huff and puff and slowly the drum would turn that wound the cable to haul the load up the incline or lower it steadily down. In fact, these early routes were not unlike the canals they served. Slowly plodding horses dragged the trucks along the level sections, just as they pulled the canal boats through the water; and where hills were met, the inclined plane and the giant, nodding steam engine replaced the long flight of locks that raised and lowered the loads.

Not all canal owners, however, regarded the tramways as simply useful feeders for their boats. Some realized that the embryo railways might grow to become dangerous rivals. The Duke of Bridgewater, known as 'The Father of Canals', looked sourly on the new system and was said to have remarked: 'We shall do well enough, if we keep clear of these damned tramways.' How right he was, though his fears were proved, albeit slowly, all too correct. His own first venture into waterways, the canal from his mines at Worsley to the heart of Manchester, was opened in 1760, and for the next forty years the potential competitor was handicapped by an all-embracing patent that effectively prevented any development of the infant steam engine. Then, in 1800, the patent expired, and the way was open for men of vision to exploit the immense power of steam. The Duke's successors would see their canal trade slip away as the rails spread their network across Britain.

It was on one of the Duke's 'damned tramways' that the first steam locomotive clanked and wheezed its way with a train of loaded wagons. Richard Trevithick was brought up in the mining region of Cornwall, where the steam engine was a vital part of the everyday life of the mines, whether pumping water from the depths or winding men and materials up and down the shaft. The Cornish engines were massive and powerful, but they were rooted to the spot. Trevithick saw no reason why, with the use of high-pressure steam, the engine should not move itself: the stationary engine could become the locomotive engine. His first experiments were with road engines and they might have been more successful had Trevithick and his friends not been rather too fond of their ale. The little engine was left simmering outside the pub where its successful first run was being celebrated: the simmering turned to boiling and the boiling became ever more furious until with one almighty bang the road engine distributed itself over a large part of the Cornish countryside.

Trevithick, however, continued his experiments, and in 1804 the two essential elements of the railway system came together: the steam locomotive and iron rails. The scene was the Penydarren Tramway in

South Wales, which linked ironworks at Merthyr Tydfil to the canal at Abercynon. The aim was to pull a 10-ton load along the 9½-mile tramway. 'It ran uphill and down with great ease and was very manageable', wrote Trevithick after the first successful run, adding a little later, '[it] is much more manageable than horses.'[2] Trevithick was, understandably perhaps, somewhat overenthusiastic in his description. Uphill and down it may have gone, but with an average gradient of 1 in 145 this was not too demanding a task. The engine was a most curious affair to modern eyes, with a huge fly-wheel on one side and moving parts that clanked to and fro over the head of the fireman who had to crouch low to avoid being brained. It was, in its way, a triumph but a flawed one. The progress of the little engine was accompanied by the harsh cracking sound of breaking rails as the brittle cast iron fractured under the weight of the locomotive.

Trevithick had done what he set out to do – he had proved that a smooth-wheeled locomotive could haul a load over smooth iron rails, but no one it seemed was interested. Partly, this was Trevithick's own fault. He received orders from the collieries of the north-east, but the first locomotive he sent was so badly constructed that it was taken off its wheels and spent the rest of its working life as a stationary engine, driving machinery. There were problems too with running a heavy locomotive on the brittle cast-iron rails then in use. Those who might have benefited from the new invention looked at the problems, shook their heads, and returned to the old, well-tried ways. Certainly there was no rush to develop the idea any further, until the Napoleonic Wars sent the price of fodder for horses shooting ever higher. In 1812, a curious locomotive set off along the railway from Middleton Colliery near Leeds to the River Aire – unlike the Trevithick engine, it worked by a rack and pinion system, which at least solved the problem of cracking rails. The rack and pinion gave extra pulling power, so that the locomotive itself could be quite small, yet still do the work. It attracted a deal of attention. Grand Duke Nicholas of Russia came to view it and showed 'a curious appreciation and an expression of no small admiration'.[3] Rather more importantly, a young colliery engine-wright from Killingworth near Newcastle upon Tyne cast an interested eye over the railway. His name was George Stephenson.

Stephenson already knew Trevithick and his work, and over the next few years he was to do more than anybody to promote the steam locomotive. He was, perhaps, more fortunate than Trevithick for Britain was now ready for the railways. The canals had been pushed as far as they could go, but there were still large areas of the country – and often just those highly industrialized regions that offered the best prospects of profit – which water transport could not reach. Already canal schemes were dropping out of favour, and rail schemes were being mooted in their place: in the north-east a suggestion was put forward that was to have momentous consequences.

In September 1810, at a meeting in the town hall at Stockton-on-Tees, the Recorder Leonard Raisbeck proposed the resolution 'that a committee should be appointed to inquire into the practicability and advantage of a railway or canal from Stockton, Darlington and Winston, for the more expeditious carriage of coals, lead etc.' There was, it seems, no great sense of urgency about the proceedings. It was two years before the engineer John Rennie came forward with his plans, and he was very much a man of the canal age. The timing was, to say the least, unfortunate, coinciding with an almost total collapse of the local banks. The canal scheme sat on the shelf until 1818, when it was taken down and re-examined, and a new plan tentatively proposed.

'The projected canal to Auckland seems now so very unlikely to get forward that I think it is a very favourable time to be prepared with some further calculations as to a railway which I am quite of opinion may be got forward with & brought forward with spirit at any public meeting that may be held at Darlington.'[4] A month later, Rennie produced what he called 'hasty productions', comparing the costs of railway and canal. It was to be the end of Rennie's connection with the work, for when he was asked to collaborate with another engineer in surveying the railway, he left in a huff – not surprisingly, since he was one of the great engineers of the day, accustomed as he said to public works 'many of them of infinitely greater magnitude and importance than the Darlington railway'. To many people the name 'engineer' suggests a steady, stable but probably rather dull individual, a practical man solving practical problems. One soon finds that, in the nineteenth century at least, they were more like overwrought prima donnas – quick to take offence, bridling at the merest whisper of a rival's name. So off went Rennie, and the work went instead to George Overton, providing a happy continuity in the railway story, for he had been responsible for the Penydarren Tramway on which Trevithick's first experimental runs had been made. An Act was duly passed in 1821 for 'a Railway or Tramroad', and if events had moved speedily forward, that is no doubt what would have been built – just another colliery tramroad along which horses would pull their laden wagons of coal. But that year there was a meeting between one of the railway's most enthusiastic supporters, the industrialist Edward Pease, and the engineer George Stephenson.

The result of the meeting was that Stephenson was able to convince Pease that the steam locomotive was now well enough developed to take over the running of the heavy goods traffic on the line. A new survey was made by Stephenson, helped by his young son, Robert, and on 27 September 1825, the Stockton and Darlington Railway was duly opened. Now, perhaps it could be said, the Railway Age had truly dawned. But had it? In its engineering, it still looked back to the old days of tramways and colliery lines. Stationary engines stood above the inclines, slowly hauling trucks up and down by cable. It did have a feature missing from the earlier lines – a regular passenger service – but the passengers were not rushed along by locomotives. They were carried by stage coach, no different from those seen on the roads, except that the wheels were flanged and ran on rails. It was a public railway, it was a major advance, but it was a line which looked backwards as well as forwards. But, if it did nothing else, it helped make the reputation of George Stephenson. It brought him to a line which, instead of linking provincial towns and collieries, was to join together two of the great centres of the Industrial Revolution, the port of Liverpool and the commercial capital of the Cotton Kingdom, Manchester.

There still remained, however, many who doubted the ability of the steam locomotive and argued in favour of a series of more than sixty stationary agents spread out along the line, between which the trains would be hauled by cable. Eventually the whole question of whether steam locomotives could travel over long distances at speed and hauling heavy loads was settled by a trial held at Rainhill in 1829. One engine stood out from all the others, the *Rocket*, largely designed by

Robert Stephenson. It proved that it could pull a train at a speed of 30 m.p.h., far beyond the wildest imaginings of the crowd that gathered to witness the trials. Just as importantly it proved it had power as well as speed. Leading experts of the day were convinced that some of the slopes, though modest by modern standards, would defeat any locomotive, and that whatever the outcome of the trials, stationary engines would be needed to help out along the way. The *Rocket* confounded the critics by storming up the slopes with no difficulty whatsoever.

The *Rocket*'s success was based on a variety of new ideas. Where previous locomotives had set their steam cylinders vertically, so that the power of the pistons was transmitted down to the track like a series of hammer blows, the *Rocket* had nearly horizontal cylinders working a pair of single, huge drive wheels. The most important innovation was the multi-tubular boiler. Hot gases from the fire passed through a series of tubes immersed in the boiler: a much more efficient system than the single tube previously in use or its near-relation, the U-shaped tube. A far greater heating surface was presented and the exhaust passing up the chimney increased the draught to the fire. In many ways the *Rocket* was the first modern locomotive – just as the Liverpool and Manchester was the first modern railway. The locomotive had vanquished the horse; both passengers and freight were moved by steam. The Railway Age was now, without question, under way.

CHAPTER TWO

The Promoters

Railway Shares! Railway Shares!
Hunted by Stags and Bulls and Bears –
Hunted by women – hunted by men –
Speaking and writing – voice and pen –
Claiming and coaxing – prayers and shares –
All the world mad about Railway Shares!

Illustrated London News, 1845[1]

The 1840s were the years of the Railway Mania when everyone appeared to be scrabbling for railway shares – tickets, it seemed, in a lottery in which everyone won the first prize. Before one can even begin to assess the work of the men who promoted the railways in the mania years, one has once again to step back in time – back beyond the opening of the Liverpool and Manchester, back even beyond the Stockton and Darlington to the tramway age of the early part of the century. Those were the days when the really hard pioneering work had to be done, when potential backers had to be persuaded not that one particular railway was worthy of their attention, but that any railway at all was worth the consideration of serious-minded men of business. Among the most important, influential and, ultimately, unfortunate of these early prophets crying in the transport wilderness was William James, claimed by his biographer, a far from unbiased judge, as 'The Father of Railways'.[2] It is certainly true, however, that he played an important part in the story of both the Stockton and the

Liverpool route, but his involvement in the iron way began before either was discussed, let alone begun.

James was a land agent who busied himself with mines, canals and railways – but it was with the railways that he fell in love. He published pamphlets, surveyed possible lines at his own expense and dashed around the country seeing new tramways and viewing the very few steam locomotives working in the early years of the century. His work began in the tramway age, and one example of his efforts can still be seen in the bridge that carried the Stratford and Moreton Tramway across the river at Stratford-upon-Avon. But his greatest enthusiasm was for the advancement of the steam railway. Between 1820 and 1821 he was surveying a line from Bishop's Stortford to Clayhithe Sluice near Cambridge and was not only advocating replacing the proposed canal with a railway, but was also proposing locomotives on the Middleton Colliery pattern.

> The old railroad system is more regular than canal conveyance, but by horse-labour more expensive; hence arose the application of locomotive steam-engines, by working by teeth and pincers on one side of a flanked cast iron rail; and hence also the perfection of them, by the invention of the Land-agent steam-engine [that is, an engine to be designed by Land-agent William James], and the malleable (or wrought) iron plate rails, secured to the present company (viz. W. James's) by letters patent.[3]

As with so many of James's schemes a great deal of effort was expended and a fair amount of cash was spent, but all to no avail. As yet the railway failed to oust the canal. James, however, was undeterred, and continued on his travels. Among his visits was one to Killingworth where he met George Stephenson. How much influence James had on Stephenson's thinking is uncertain, but they were certainly friendly enough for James to offer to promote Stephenson's engines and to engage young Robert Stephenson to help on a scheme which could have been his greatest triumph, the Liverpool and Manchester Railway. As usual, James undertook all the initial work himself, writing pamphlets and undertaking a rough survey of the line. He published his preliminary results in 1822 and was able to find just the man he wanted to help forward the scheme. Joseph Sanders was a successful and wealthy Liverpool merchant who, like many other local businessmen, was heartily sick of the canal companies and their abuse of monopolistic power. They offered a poor service at high prices.

> The canal proprietors were dilatory to the public until they became dangerous to themselves. Although the facilities of transit were manifestly deficient; although the barges employed to carry goods often got aground, and were sometimes wrecked by storms; although for ten days during summer the canals were closed; although in very severe winters they were frozen up for weeks; yet they established a rotation by which they sent as much or as little as suited them, and shipped it how or when they pleased.[4]

But work on the Liverpool and Manchester scheme went forward very slowly, and James was now paying the price for his railway promotions. His other concerns had been neglected and he was facing bankruptcy. In May 1824, the last and cruellest blow fell. The Liverpool and Manchester Board took the work away from him:

> I very much regret that, by delay and promises, you have forfeited the confidence of the subscribers. I cannot help it. I fear now that you will only have the *fame of being connected with the commencement of this undertaking*. If you will send me down *your plans and estimates, I will do everything for you* I can; and I believe I possess as much influence as any person. I am quite certain that the appointment of Stephenson will, under all circumstances, be agreeable to you.[5]

It was far from agreeable: it was, in fact, a double betrayal for James. It was he who had put the whole scheme to Sanders, and now his place was to be taken by the man he had befriended, and to whom he had freely offered valuable advice. James's supporters made much of the calamity, but to be fair James was by no means the only man keenly supporting railways in the 1820s – nor indeed the only one setting forth arguments for a line from Liverpool to Manchester. The railway historian John Francis, writing in mid-century, puts forward the rival claims of Thomas Gray, a man at least as obsessed as James: 'Begin where you would, on whatever subject – the weather, the news, the political movements of the day – it would not be many minutes before, with Thomas Gray, you would be enveloped with steam.' Here were 'visions of railways running all over the kingdom, conveying thousands of people and hundreds of thousands of tons of goods at a good round trot; coaches and coachmen annihilated; canals covered with duckweed; enormous fortunes made by good speculators, being talked of as sober realities that were to be.'[6]

Nevertheless, it is hard to see James as he left the north for a future of court cases, debt and penury as anything other than a tragic victim of his own enthusiasm. His one consolation could have been that when he began his campaign the world was at best indifferent, and at worst hostile, to the very idea of steam railways: with the opening in 1830 of the Liverpool and Manchester Railway which he had so ardently promoted, the steam railway was accepted as a system of proven value.

Although James put much – too much, indeed – of his own money into promoting railways, their success in the early days depended still on men of wealth and standing giving their practical support. Returning again to the early years, one finds Edward Pease, industrialist and merchant of Darlington and principal promoter of the Stockton and Darlington Railway. As a Quaker he was barred from public office, but this did not prevent him from acting in the public interest. By the time of the railway promotion he was in his fifties, prosperous and able to act without any thought of direct payment – happy to settle for the long-term benefits that the line would bring:

> My favoured position in life did not render any remuneration for service needful, nor did I ever receive a shilling, or dinner, or anything for my exertions. When towards the close of our work, money falling short, our banker refused to grant us more, I paid all the workmen, &c., &c., employed in this way out of my own resources, until he could procure a loan. I remained in the direction one year after opening, until 1827, when seeing there was

> income enough to pay a handsome dividend, I retired with a resolution never to enter a railway meeting again![7]

As a generalization it can be said that the early railway schemes were promoted by businessmen, merchants and industrialists who, like Pease, looked to long-term gains rather than short-term profits. Shares in these early schemes were not cheap. Those for the London and Birmingham, begun in 1833, were available at £10 or £25, and the list of subscribers shows many investors holding hundreds of shares.[8] These were men of substance, listed as merchants, calico printers, manufacturers and the like. Interestingly, in the list is one Bartholomew Bretherton of Liverpool with £2,800 invested and listed as coach proprietor, a gentleman who was presumably working on the principle 'if you can't beat them join them'. It is also notable that a very great proportion of the money was raised in Lancashire, where men had already seen for themselves the benefits that a railway could bring. It was not always so easy to convince others. A similar story can be told about the start of the Great Western Railway, which was said to have had its origins in a meeting in 'a small office in Temple Backs'[9] in Bristol: the promotion was spearheaded by a group of four businessmen who between them were to invest over £60,000 in the new concern. Their enthusiasm was not, however, matched by that of their fellow Bristolians and much of the money was raised in London.

New projects were always floated on a sea of optimism:

> A new œra has however arrived, and the intelligent inhabitants of the West Riding of Yorkshire have been among the first to hail its advent, the extension and perfection of the system of internal communication had too frequently occupied their attention, and their isolated situation had too frequently been the subject of their regret, to permit them to regard with apathy or indifference the grand and interesting experiments on the capabilities of the Railway system, and on the practical application of the powerful agency of Steam to the purposes of locomotion.[10]

There was an abundance of purple prose in this enthusiastic view on the advantage of rail links to Sheffield, and a marked reluctance to dwell on the problems of pushing railways through the Pennine hills. Equally fine words came from the speakers at public meetings held to promote the new railways.

In 1835 a variety of learned and noble gentlemen were being trundled around East Anglia to tell of the delights and profits of the proposed Eastern Counties Railway. This was to be a major undertaking, 125 miles long, but if its promoters were to be believed, all could be achieved with the minimum of expense and trouble, and in the end would yield the maximum of profit. The engineer came forward to declare that as the land was so level the costs would be correspondingly low. He added, for an unexpected reason: 'it has this great feature – that it has no tunnels'. This was a reasonable comment given the notoriously unpredictable costs of tunnelling, but in fact in this instance the engineer had something else in mind. 'Tunnels are intolerable nuisances, which sooner or later must prove fatal to any railway that has them. It is impossible for any person who has travelled once in a tunnel to do so for a second time.'[11] Other claims were even more dubious. He went on to declare that the line 'interferes with no ornamental property or houses' – a statement which was not intended to include a large swathe of London's East End where he calmly announced they would be 'obliged to sweep down many – but they are poor houses'. But no great houses? The meeting was not told about the deal that was being put to Lord Petre whose land was needed. The company offered to pay him the huge sum of £120,000 in compensation – not that the land was worth a fraction of that amount, but this would ensure that His Lordship did not oppose them in Parliament. When the bill was passed, the company tried to wriggle out of the deal, but in return '[Lord Petre] opposed and prevented them from passing through his grounds; he harrassed and irritated them as they had irritated him, and with far more effect.'[12] He got his cash. The question of fighting off the opposition was also raised, and firmly dealt with: 'It is the determination of the Committee not to spend a shilling in Parliamentary warfare'. In the event parliamentary expenses came out at £35,000. One after another speaker rose at public meetings to eulogize the line: one felt that, thanks to railways, he would 'live to see misery almost banished from the earth'; another, more mundanely, looked forward to 'enormous profits for the shareholders'. A few querulous doubters were assured there would be no harm done to local roads and waterways, and the whole proposal was greeted at the end with virtually unanimous approval. Verbal support was forthcoming, but not cash, or at least not locally. Shares were widely advertised, in the press and in specialist newspapers such as *The Railway Times*, and were available through brokers. On this line, only a twelfth of the capital was raised locally. Railways were beginning to take on a different character, and the Eastern Counties, begun with such rapturous promise, was destined to play an altogether less happy role. It was soon to fall under the sway of the man who was to dominate and personify the years of the Railway Mania, George Hudson, the Railway King.

The distrust of railways that characterized the early years of the 1820s gave way in the 1830s to a rising enthusiasm that was to become divorced from rational discussion. Where once it had been fashionable to decry railways as mere fads, it now became equally fashionable to cry them up as miracle-workers, ushering in a new age of prosperity. Unfortunately for the development of a rational system, there was no controlling hand at work. Anyone and everyone was free to issue prospectuses and to raise capital for schemes, whether wholly worthy or totally impractical. Railway scheme competed with railway scheme, and all too often a new line would be proposed with little if any consideration of how it might fit into the overall pattern. As the system grew, so the case for rationalization grew stronger. One solution would have been for the government to step in, but in the great age of *laissez-faire* politics this was never really a practical alternative. So the way was open to any man of vision who could command the necessary resources to take on the task for himself.

Hudson and the Railway Mania are in many ways inseparable: one scarcely makes sense without the other. The mania grew from an artificially raised and ultimately falsely optimistic view of railway profitability; at the same time the workings of the system of promotion made it all too easy for the unscrupulous to make exorbitant profits. The effects of the mania were plain for all to see:

> You saw a man to-day in the streets of Bristol whom you would not trust with the loan of a five-pound note; to-morrow he

> splashed you with the wheels of a new Long-acre carriage. He was as suddenly transformed from a twenty pound house to a mansion in the country, and though small beer refreshed him during the greater part of his life, he now became critical in the taste of Bordeaux.[13]

This is all a far cry from the devotion to the public good of a man like Pease. How did it come about? The Eastern Counties provides a useful case history, but first one has to look at the career of George Hudson.[14]

He was in every sense an overbearing personality. A big man physically with a voice to match, he was also a man of outsize ambition. His start was modest: a farmer's son who became a linen draper, but in 1887 he inherited £30,000, became involved in banking and thrust himself into local politics in York. As an early enthusiast for railways he was to form a close working relationship with George Stephenson. The Stephenson connection was of immense value to Hudson. Time and again he brought the engineer along to public meetings, whether held to promote new lines or shareholders' meetings of lines already begun, and such was the esteem in which Stephenson was held that no one thought to question Hudson's honesty either. Yet his, to put it kindly, amoral attitude to finance appeared early on when he became Treasurer for the local Tories and was hauled up before a Commons committee in 1835 to explain some of the more dubious expenses. Since venality was common in those days – and as prevalent among the Whigs as the Tories – no more was heard of the charges. Soon he was heavily involved in railway promotion, working on schemes designed to make York the premier railway city of the north, a vital link in the line from London to Scotland.

Railway development in Scotland had been proceeding piecemeal throughout the 1830s, and it was obvious that at some point rails would unite Scotland and England. It was equally clear that, if the border hills were to be avoided, the main-line routes would have to go up the coastal plain, either to east or west. The vital question seemed to be: who could first pull the disparate elements together to create a genuine trunk route from London. Hudson's base at York was already connected to the English Midlands by two railways, the North

Midland and the York and North Midland, both opened in 1840, and he was determined that they should lie at the heart of the network. He also understood that this would only happen if control could be centralized. He began steadily taking over smaller, often struggling, companies in the name of the York and North Midland. In view of later events, it is easy to characterize George Hudson as an out-and-out villain, but his contribution to railway development was both real and necessary. In 1844 he created the Midland Railway by amalgamating three other successful companies, with nearly 200 miles of line serving the busy industrial communities of the Midland region. This was the positive side of his achievement, and he received the enthusiastic backing of the shareholders, encouraged by the high dividends paid by the Hudson companies. His motto was 'every shareholder an auditor', and he ensured that the shareholders backed him by regularly turning up at the annual meetings and demolishing any opposition by the power of his rhetoric and by smothering argument with a welter of usually unverifiable statistics. When one company had the temerity to appoint a committee of enquiry, they arrived at the first meeting to find the king enthroned. He had been empowered, they were told, to fix the date of the first meeting and they would be informed when that was to be. That was the first and last meeting of the committee.

He was equally brusque with the directors of the Midland. Hudson, as was his wont, had made a deal in the company's name without even the pretence of consultation. The directors gathered, somewhat apprehensively, one imagines, at the next board meeting, as Hudson strolled in.

> 'How, now, gentlemen,' said Mr Hudson, 'has anything happened?'
>
> 'Only', replied one, 'that we being equally responsible with yourself, for what is done, are desirous of knowing the nature of your plans.'
>
> 'You are, are you?' rejoined the railway monarch, 'Then you will not.'
>
> And the business of the board proceeded.[15]

There had always been doubters, questioning Hudson's apparent ability to pay higher dividends than other companies, just as there were those who questioned the vast personal wealth the Railway King was accumulating. The Eastern Counties Railway provides a perfect example of the system at work. It was not, by any reasonable standards, a success. The line, so confidently promoted in 1836 as the grandest in the kingdom, had by 1843 only reached as far as Colchester and the cash had run out. It was a line of remarkable follies and considerable inefficiency. The country was then divided between the proponents of Stephenson's 'standard gauge' of 4 ft. 8½ in. and the Brunel broad gauge of 7 ft.: each certainly had its merits. It is, however, difficult to see what conceivable advantage the Eastern Counties could have hoped to achieve by turning its back on both and sticking an extra 3½ in. on the standard gauge and setting its rails 5 ft. apart. Inevitably, they all had to be pulled up and relaid. Even as late as 1856, a local man was putting an advert in the papers offering to race his donkey against the railway:

> He will do what I have stated with ease, and have a good bray afterwards, as if in contempt of the inferior power of Eastern Counties Steam.[16]

Perhaps wisely the company refused the challenge. Yet this was the company which, as soon as Hudson took over in 1845, began paying out absurdly high dividends. The questions could not remain unanswered for ever. Once the books were properly inspected, the nature of the business became all too clear. The books had been well and truly cooked. The apparent profit of £545,714 shown in 1845 was created by moving cash from other accounts – £320,572 of it. The calculation of the profits should have been straightforward: basically, take the revenue earned by carrying passengers and freight, deduct from it the running expenses of the line, such as salaries, fuel for the engine, maintenance, repairs and so forth, and what you were left with was the profit. Hudson could do nothing about the revenue side, but he could do something about the costs, by transferring these to the capital account. This was supposed to be limited to money spent on capital projects, such as a new bridge or station building.

Hudson set about hiding basic running costs in this other account. The result was that running costs appeared remarkably low – falsely low – and the profits correspondingly high. The shareholders were, at first, quite happy with this arrangement, which resulted in high dividends, and big, bluff George Hudson was always on hand at their meetings to quell any doubters. They failed to notice that among the shareholders who were milking the company was Hudson himself. The facts inevitably came to light: Hudson had personally taken away the books and altered them. The shareholders were, not surprisingly, angry, and Hudson for once failed to appear at their meeting and left his deputy chairman David Waddington to face the music. The meeting was held on 28 February 1849 and Waddington was lost before he began.

> I stand here in a painful position – ('No doubt you do') – I say, it is most painful to think that one with whom I was formerly on the most intimate terms of brotherly friendship ('Oh, Oh' and laughter) – it is painful for me, I say – (Groans, hisses and cries of 'Sit down, sit down').[17]

Only one statement brought a cheer, the announcement that the Board of Directors had resigned. This was not enough for the shareholders, who demanded that Hudson and Waddington be brought to trial. The fall of George Hudson was sudden, precipitous and final, and the railway world that had fêted him turned on him now with savagery. The press trumpeted their disapproval.

> We thus have the Chairman and Board of Directors . . . convicted of grossly and palpably falsifying the accounts of the Company, certifying to their ability to afford an amount of dividend not earned by the traffic; we have expenses incurred for which no reason can be assigned – we have payments made of which no records exist – we have accounts cooked (a pretty hash) by one Director, against the protest – (we must in justice state) of his only operative colleague. We have, in fact, every vice which can taint the system of railway management.[18]

Satirists remembered Hudson's shopkeeping past.

> Good accounts are troublesome things to keep, and occasionally cause trouble to the parties of whose affairs they are register. The true chandler's shop system is to keep no books at all. A cross for a halfpenny, a down stroke for a penny, a little O for a sixpence, and a larger for a shilling, all in chalk, on a board or a cupboard door, constitute the accounts of many a money-getting shopkeeper; and, we doubt not, would suit well the purposes of some of the railways. Chalk is easily rubbed out and put in again; ink is a permanent nuisance.[19]

It is easy to vilify Hudson. He was an undoubted villain who pocketed money that was not his, manipulated shareholders and skilfully – but not quite skilfully enough – covered his tracks by creative accounting. The account books of George Hudson and his companions must rank among the masterpieces of nineteenth-century fiction. Nothing quite like it was to be seen again in British public life until the advent of Robert Maxwell. But, given the response of the British government to the Railway Age, which was simply to stand back and view the dealings with patrician disdain, perhaps the railway world needed such a giant ego. The system was, after all, to be the major trade route of the country for a century and more, and it was developing with no consideration for overall needs nor to any coherent pattern. Hudson's motives may have been – were, indeed – dubious, but his realization that someone had to bring order to the chaos was crucial for rational development. The Midland Railway was the first major company to be formed through amalgamations and was to become one of the great railway companies of Britain: its locomotives in their distinctive livery of 'Derby Red' could be seen steaming out from the High Victorian splendour of St Pancras *en route* to Glasgow or Edinburgh. Whatever his personal faults, Hudson left behind a proud and successful railway company, though that was not necessarily a comfort to the thousands who lost their savings and were faced only with penury and ruin.

The excessive and fraudulent dividends paid by Hudson fed the fires of Railway Mania. If even so wretched a concern as the Eastern

Counties could show a handsome return, why should any railway fail? One commentator, at least, was able to take a more balanced view of the whole affair, seeing the Hudson fiasco as the product of a general malaise of greed and stupidity.[20]

> Mr Hudson is neither better nor worse than the morality of 1845. He rose to wealth and importance at an immoral period; he was the creature of an immoral system; he was wafted into fortune upon the wave of a popular mania; he was elevated into the dictatorship of railway speculation in an unwholesome ferment of popular cupidity, pervading all ranks and conditions of men.

What was the system that allowed for such a chaos of greedy dealing? Parliament, anxious to ensure that railway promotions were viable, demanded as a first step that 10 per cent of the estimated cost of the line be put on deposit and that a list of subscribers be presented promising a minimum of three-quarters of the whole capital needed. It was with the subscription list that many of the problems started. Anyone signing as a subscriber received scrip, a document entitling them to buy shares at face value as soon as they were issued. In the general euphoria surrounding railways there was a firm belief that as soon as the shares were issued they would increase in value – a similar phenomenon could be seen at work in recent years when the British government began selling off public utilities. There was a general expectation of initial profit. So it was in the Railway Mania years, when speculators had only to sign a subscription list to get their hands on scrip which they would instantly sell on. Those men, the stags, were desperate to get in at the start of a scheme. At a meeting to launch the Runcorn Railway, some would-be subscribers 'who could not get near to sign their names soon enough, crept *under* the table, and emerging between the knees of those who surrounded it thus secured the chance of having their names enrolled'.[21] No part of Britain was immune: the following complaint came from Blackburn.

> There was a time when speculating in railway shares was left in this neighbourhood to the working of cool heads and full pockets. Now-a-day, however, so rapid has been the intellectual

> progression of the times (it is to be lamented that the march of money has not kept time with the march of mind,) that anybody literally anybody or as it may perhaps be more appropriately worded literally nobody rushes into the market and with more brass in his face than in his pocket by a good deal straightway becomes a dabbler in 'London and York's', 'Dublin and Belfasts', 'Edinburgh and Glasgows', nay, even 'Paris and Lyons' or 'Strasbourg and Basle's!' Of course those schemes nearer home come in for an extra share of patronage; and amateur brokers block the way with 'I say, have you got any Bolton's and Darwen's?' 'How are you off for Accrington's?' 'What's the state of the Ormskirk's?' 'What'll you take for Preston's?' Such is the excitement at present among some of the speculators in Blackburn that having obtained possession somehow or the other of an 'allotment' they pounce upon their unhappy victims at street corners, shop doors, public-houses, before, during and after business anywhere exclaiming 'Here my good fellow, here's a chance for you; you may have these Bolton's at 5l.11s, well, say 5l 10s 10d: – well, then, make it 5l 10s. 9d., just threepence a share more than I gave for them!' It is in vain that an escape is effected in one place; you are watched for and seized upon at another with an offer of 'Ormskirk's' 'West Ridings' 'Prestons' 'Rugbys' any of them at any sum above prime cost, so that a profit may be made.[22]

Even with this rush of subscribers, companies were often desperate to find enough names to fill the list for presentation to Parliament, and were often reduced to persuading anyone to sign up for shares, regardless of whether or not they could pay for them. If even that failed, they invented wholly fictitious subscribers. Such devices would have gone unnoticed, for Parliament was deluged with railway business and had neither the means nor the inclination to check the lists, but rivals did their work for them. In among the rush of prospectuses, there were some who put up schemes which they knew full well would never be built, which they never intended to see built. They took the speculators' money and, after a suitable amount of time had elapsed, announced that their plans had failed and alas the money was spent, though there was an understandable reluctance to explain just where it had gone. Even railway companies themselves were guilty of wild speculation. The directors of the Caledonian, having raised funds for the construction of the line, invested £380,000 of this capital in shares in other companies. When the mania ended and prices tumbled, they found they now had no money left for their own railway.[23]

What many speculators conveniently forgot – assuming that they ever knew and understood – was that railway shares carried responsibilities as well as profits. The companies were empowered to make calls on shareholders as more cash was needed – and getting rid of the shares was no easy solution. It was often, indeed, impossible. Nor could debts be written off that easily. The bubble, inevitably, burst, but in the meantime a vast amount of money went into railway plans. Investment in English railways stood at a modest £170,000 in 1825; by 1844 it had risen to over £67 million, overtaking even such great concerns as the booming cotton industry.[24] The pattern for the period is certainly a mixed one; statistics show just how mixed. A survey carried out in 1853 revealed that up to that time nearly £190 million had been raised in share issues and another £57 million raised by loans, as a result of which 7,686 miles of railway had been opened. But set against that rosy picture was the fact that in one year, no fewer than ninety-five companies had gone to the wall.[25]

After the mania, a certain sobriety came into procedures. In the 1850s it was no longer enough to have a railway with a good name and a chairman with an equally imposing title:

> You may plant at the head of a Company a prince of Royal blood, if you please, but the concern, in its aims and details is not thereby ennobled . . . Gentlemen who never cast their eyes over a tradesman's bill, nor care what amount they draw for at their bankers, are among those least qualified for the duties of Railway Directors. Everything in these duties must be of detail. With what grace can you ask a man to investigate or explain the variations in the prices of coke and grease, who knows nothing of the cost of the one, and faints at the smell of the other?[26]

A new attitude appeared among investors. The Eden Valley Railway approached a potential investor, but when he heard that the leading proprietors were prepared to put no more than £500 of their own money into the project he sent them a decidedly dusty answer:

> Now if this is the case & the leading parties do no more where is the money to come from? I see no likelihood of getting it elsewhere if the managers have so little pecuniary interest . . . They have nothing at stake who manage it. Now unless you see a clearer way in the mist than I do, I should *not* like to have anything to do with it, merely to waste money in a fruitless attempt . . . Neither Mr Pearce nor I ought to be interested in a new bubble to be blown in the air which we knew beforehand could cost us £100 each and fall flat when burst.[27]

There was good evidence that this is precisely what had happened in the past. The modest Maldon, Witham and Braintree Railway, begun in 1846, raised £45,500 in London, a mere £11,360 in Maldon, £24,840 in Witham and nothing whatsoever in Braintree. Within a year the line had been eaten up by George Hudson and the Eastern Counties.[28] Later branch lines were to be much more dependent on local, and interested, support. At a meeting to promote the line from the Great Western to High Wycombe, investors ranged from a Thame plumber buying one £10 share to 'a gentleman' putting his name down for 100. The largest item, however, shows another aspect of the age: £35,000 subscribed by two contractors, Thomas and Solomon Treadwell. Contractors wanted the work, the locals wanted the line. Together, they could be reasonably sure of getting it.[29]

In 1849, *The Railway Times* could declare in its editorial pages that the 'delirium of expectation' was over. The days when share-brokers made easy fortunes were gone, and shares sold at a reasonable price could be expected to show a reasonable return. Accounting had improved, and the possibility of Hudsonian-like fraud had diminished. True, money was now short, but the future looked bright. The old days when sound men of business promoted sound railways were about to return: the evil days were gone for ever. Or so it seemed to commentators at the time. Industrialists promoted lines to serve their concerns; local communities fought to join the network, so that they too could enjoy the benefits of fast, cheap travel. A town without a railway was, it seemed, doomed to a future of stagnation and decay. The railways brought new communities to life: the fishing hamlet with a railway could become a prosperous holiday resort; its neighbour along the coast without one would remain a huddle of cottages round a crumbling harbour. All this seemed a far cry from the bad time of the mania years. Perhaps so, but it is easy to lose sight of the fact that in the mania years good, profitable lines were promoted alongside the failures, and that vast capital was raised and much of it was actually spent in building railways that were of real benefit to the communities they served. It was not all loss. But whatever the nature of the line in which the optimistic raced to invest, there was a very long way to go before the line was built and the first train ran.

CHAPTER THREE

Preparing the Plans

The landed proprietor often refused admission to the trespasser and to his theodolite. At Addington the surveyors were met and defied in such force, that after a brief fight they were secured, carried before a magistrate, and fined . . . The engineers were in truth driven to adopt whatever method might occur to them. While the people were at church; while the villager took his rustic meal; with dark lanthorns during the dark hours; by force, by fraud, by any and every mode they could devise, they carried the object which they felt to be necessary but knew to be wrong.

John Francis, *A History of the English Railroads,* 1851

Having rallied support, acquired some fine-sounding names for the letter-head and received promises, however dubious, from investors, the railway company now had to start preparing plans for Parliament. The government had to decide what plans were needed and, inevitably, set up a committee for the purpose. The parliamentarians were somewhat concerned about the scale to which plans should be drawn – would six inches to the mile prove troublesome? Their first witness, Colonel Fanshawe, replied with true military brusqueness that if the work had been done properly there would be no difficulty, and if it had not been done properly then they should not bring any plans forward at all. Another engineering witness proposed a larger scale for villages and towns than for the open countryside, pointing out that a recent bill for a line into London was accompanied by a 3$^{1}/_{10}$ in. to the mile map on which it was impossible to see what property would be affected. Then came the difficult problem of fixing a date by which plans had to be deposited for consideration in the next session. James Walker, a civil engineer, pointed out that the best time for surveying was during the period immediately after the harvest was brought in, and that engineers too had their seasons:

> Engineers and surveyors of any practice are fully employed in reaping their own harvest here in the House of Commons, during the sitting of Parliament . . . I have been here in town for two months, without being able to get out of town.

When asked why he could not leave the work to his assistant, Walker replied that they were too busy making changes to existing plans.

> Now yesterday there was an objection to a railway I was concerned in; an owner thought he had found a better line, and nothing would satisfy him but I must either send or go down myself to examine his line, or if I did not he would start an opposition in the House of Lords against the measure.

In the course of all this detailed discussion, an idea was slipped in that the various companies should put up plans for local lines, but that the government should decide on track routes, to ensure the country got a rational, nationwide network. The work could then be put out to tender. How very different railway history – and the railway system – might have been if that suggestion had been taken up. Instead the committee restricted itself to plans and dates. The last date for lodging plans was eventually fixed at 30 November, a date which was to be engraved for ever on many a harassed engineer and surveyor's heart.[1]

In theory the parliamentary survey was a straightforward, if laborious, exercise. William James was one of the first engineers to set it out in clear, lucid steps.

> The first step is to ascertain the amount of tonnage likely to pass the whole line or partially, the nature of the trade, and the speed indispensable for its transit.

> The next step is to discover the best line, I mean the shortest and flattest for this purpose. A section and survey of investigation must be taken by an *engineer*, (not surveyor under his direction, but by himself – for in case error is afterwards found out, *he* will escape responsibility, by blaming the surveyors.) This preliminary survey will afford an estimate or approximation for the requisite capital, to enable you to fix the prospect of profit and loss so as to see the prudence of perseverance.[2]

This was all very well in theory but could be more troublesome in practice. One manual gave advice on how traffic could best be estimated.[3] Men had to be sent out to the termini and towns along the way to count all the traffic on roads, canals and rivers. Then the engineer had to work out how much extra traffic the railway could expect to bring in. For example, a traveller in Bedford would be better off going by coach to the London and Birmingham Railway and continuing the journey by rail as he would save one hour and nine minutes, but a would-be traveller in Hatfield would be better off going all the way by coach. This would seem to have been the first attempt to lay down rules for railway commuters. The thankless task of counting country coaches and post-chaises, wagons and carts usually went to some hapless junior engineer who spent long hours out in all weathers noting down everything from Farmer Giles' wagon to the mail coach from London.

Finding the best line was far more difficult. Once a railway company had its Act of Parliament, the surveyors and engineers could march with impunity over the land required for the route, but until then they had to rely on the goodwill of the landowners. Brunel gave a little lecture on the need for tact and civility to his surveyors, though he himself was not notably endowed with either.

> In conducting this portion of the survey of the Great Western Railway entrusted to you it is particularly desirable that you should ascertain the names of the owners and occupiers of any land to be passed over and as early as possible obtain their sanction to the making [of] the survey. It will of course very often be impracticable to obtain this information until you have

actually entered upon the land but if you make the application as early as you can I think you will not find any vexatious opposition thrown in your way. You will at all times afford every information to persons interested in the property likely to be affected by the Railway and you will consult them as to the manner in which the line may be carried so as to be most advantageous to their property explaining to them at the same time if necessary any circumstances which from the nature of a Railway may render such a distinction difficult or impracticable in order that as far as consistent with the absolute requisites of a public work of this description the line ultimately selected may be adapted to existing interests.

You will be particularly careful that all persons employed by you shall conduct themselves with propriety and civility that no damage be done to the hedges or fences and that they do not unnecessarily encroach upon gardens or enclosed ground or otherwise annoy the inhabitants.[4]

This was all very well, but it assumed goodwill on the part of the locals which was not always present, certainly not in the early days of railway promotion. What were the engineer and his surveyors to do when a landowner refused access and resolutely barred the way? William James found his own calm advice of little help in practice when he came to survey the route for the Liverpool and Manchester. He was faced with local villagers full of dread of the fire-eating monster that would invade their fields and with landowners – especially those with shares in the old canals of the area – who even went so far as to threaten physical violence. George Stephenson, when he came to survey the same route, had an equally difficult time:

We have sad work with Lord Derby, Lord Sefton and Bradshaw the great Canal Proprietor whose ground we go through with the projected railway. Their Ground is blockaded on every side to prevent us getting on with the Survey – Bradshaw fires guns through his ground in the course of the night to prevent the surveyors coming on in the dark – We are to have a grand field-day next week, the Liverpool Railway Company are determined to force a survey through if possible – Lord Sefton says he will have a hundred men against us – the company thinks those great men have no right to stop a survey.[5]

The engineer fought back with his own stratagems.

But one midnight night a survey was obtained by the following ruse. Some men, under the orders of the surveying party, were set to fire off guns in a particular quarter; on which all the gamekeepers on the watch made off in that direction, and they were drawn away to such a distance in pursuit of the supposed poachers, as to enable a rapid survey to be made during their absence.[6]

Others were obliged to be even more devious. The Duke of Cleveland, whose lament at the prospect of a railway running past his lands has already been heard, at once ordered all his employees to keep out the surveyors at all costs and by any means.

The Duke of Cleveland has issued the strictest orders to all his tenants and servants, on no account to allow any railway engineer to make a survey through any part of his property, by giving all who attempt to do so notice to desist; and if this be disregarded commencing immediately an action for trespass against them. A party of surveyors recently started off from Barnard Castle about 2 o'clock in the morning, thinking that they should be thus enabled to accomplish their object; but they were mistaken, the watchers were wide awake, and fully prepared to hinder their progress and the discomfited engineers were obliged to return. Another party surveying in a different locality near Barnard Castle met with a similar reception, watchers being posted night and day.[7]

The company tried a different approach, earning a stern rebuke from the Duke.

> You have attempted to make a survey near my property by stealth, and to make the fraud more complete you ordered your surveyor to put on the dress of one of the Gaffers and Miners, pretending that he was taking a Government survey.[8]

It was all very trying, but sometimes a little humour breaks through – even if it was not appreciated by everyone taking part. John Fowler, later Sir John Fowler, eminent engineer, began in a modest way as a young engineer helping to take surveys. On one outing when they were working on the public road, the servant of one of the local landowners plonked himself down, declaring that as it was a public road he as a member of the public had a perfect right to stand anywhere he chose.

> On this, one of my stalwart assistants inquired if he also had the right to stand or walk where he pleased on a public road. Unthinkingly the landowner's man admitted that any man had such legal right. 'Then', said my man, 'my right is here, and if you obstruct me I shall remove you'; and walking up to the man, he took him in his arms and deposited him in a ditch.[9]

Things, however, could take an altogether more serious turn. In 1845, the Midland Railway wanted to push a line from Leicester to Peterborough through Lord Harborough's land at Stapleton Park. But His Lordship had just spent a lot of money on improvements to the park, and he had no wish to see it ruined by a railway. The company thought it had found the solution by ordering its surveyors to work from the towpath of the Oakham Canal. Unfortunately, Lord Harborough had rights over that as well – illegally, as it turned out, but no one was worrying too much about niceties. The surveyors were stopped by a group of men headed by one of the estate keepers. Things began to take an unpleasant turn when one of the railway men pulled a pistol, but the keeper called his bluff and after that there was nothing to do but retreat. The result of the first expedition was no surveying completed and a weekend in gaol for the gun-toting surveyor.

The next expedition relied on brute force. The railway men led a gang of navvies to Saxby Bridge where they were confronted by an equally determined band of estate workers. By now, however, the local police had become aware of the dangers of the situation and were able to keep the two sides apart. There were a few scuffles before the leaders of both sides agreed to meet and argue their cases before the magistrates. The police went away satisfied. However, Lord Harborough's men did not trust the railway men – rightly as it turned out. At dawn nearly a hundred railway men crept up on the site, but the opposition was ready for them. This time there was a thoroughgoing battle with staves and cudgels. Astonishingly, while the fight went on the survey was somehow completed, though one would love to know how accurate it turned out to be. Six of the railway gang were sentenced to a month in prison, which they found rather easier than navvying, but as was often the case, it was the gentry who eventually emerged as winners. The line was built with an extravagant sweep round the edge of the estate, which came to be known as Lord Harborough's Curve.[10]

Freed from interruptions, the job of the surveyors was one that required a certain expertise, a hardy constitution and a great capacity for slow, careful work. The railway engineers and surveyors were more fortunate than their predecessors of the canal age, for in 1791 the Ordnance Survey was founded and promptly set about preparing accurate maps of the whole country. Armed with such a map an engineer could set out the rough line the railway should follow. 'Flying levels' were taken, using a simple mountain barometer to estimate altitudes, and all that was needed was the energy to walk the route. Robert Stephenson is said to have walked the ground between London and Birmingham twenty times before deciding on the best route. The detailed surveys required altogether more accurate measurements. The equipment was simple enough – theodolite, graduated staffs and chains for measuring distance. The 'chain' measurement of 22 yards has virtually disappeared now, surviving only on the cricket pitch as the distance between wickets, but for the nineteenth-century surveyor it formed the basis of all his work.

The chain was both a measurement and an actual physical object, a chain of metal links being used in the field. Experienced surveyors became so used to this device that it was said they could estimate the distance to within a fraction of an inch by eye. Straight lines were simple, but curves presented more of a problem. Practical handbooks full of impressive mathematical equations offered help and guidance, if not always with perfect clarity.

> Some of them have given an unnecessary profusion of formulae, which involve the subject they pretend to explain, in such a degree of obscurity as must be very perplexing to students, and repulsive to those engineers who have been accustomed to use my methods.[11]

But mathematics was no substitute for common sense. One survey on the Wisbech, St Ives and Cambridge Railway had simply taken the shortest route at one point, cutting across two roads, each requiring expensive crossings, going through three gardens and involving the destruction of three cottages. A diversion of a few yards to a point just past where the two roads met would have involved just the one road crossing and would have missed houses and gardens completely. It was, as the commentator wryly noted, an expensive blunder, 'but these matters are often strangely overlooked, or disregarded in the hurry of railway practice'.[12] Some of the best advice for surveyors came from practical engineers.

> There are a number of conventional rules without which the results are unintelligible except to the persons who took the levels – it is necessary to commence at some known level – to establish good bench marks both there and as frequently as possible along the road – having one general rule for selecting such marks as for instance the tops of milestones – the sills of doors – and the top of the lower staple pins of gates – the levels should be carried as much as possible in short straight lines from points that can be determined upon the ordnance maps – always passing if possible the lowest summits and lowest parts of hollows – taking note at the same time of the rise or fall on each side.
>
> 4 or 5 levellers will be needed, but in trial levels pacing out distances gives a good enough approximation. A 4 mile to the inch horizontal and 40 ft vertical is a recommended scale.[13]

Surveying was often entrusted to young men just setting out on their engineering careers. F.R. Conder was a sub-engineer on the Eastern Counties when he was still 'a young man not out of his articles and an Engineering pupil'.[14] He had to recruit and train staff-holders and chain men, and when he did set off he found that the chief engineer's instructions were, to say the least, vague and the maps wholly inadequate. Then, to make matters worse, the terrain was appalling as he squelched through the bogs of the Fens:

> It was only possible to obtain accurate levels across this morass by spreading the legs of the instrument to their widest possible extent, and kneeling between them to observe – a pleasant occupation for a damp autumn day.

His assistants were mainly recruited from amongst former army men, good, solid, dependable ex-NCOs. But they did include in their ranks an Irishman called Dempsey – 'very deaf, very stupid, very ugly and indomitably conceited', whose sole virtues were absolute honesty and sobriety. He did, however, suffer from what he mournfully referred to as an 'incurable intarnal disease'. It was some time before anyone discovered that the sobriety was a myth and the incurable disease was poured from a bottle.

Conder's ambition was to become an engineer, but others were quite content to remain as surveyors, an occupation which promised a very reasonable standard of living. One surveyor, William Fairbank, had quite enough work on hand to be able to employ George Hamilton, who termed himself rather grandly 'Civil Engineer and Architect'. Hamilton set out his terms for conducting the parliamentary survey for the Chesterfield, Sheffield and Gainsborough Railway in 1844:

> The terms will be two Guineas per day for each day I am from home on the said business, and all reasonable personal expenses & 5/-per day for Staff-holder to include all expenses except coach hire and wages of Chain leader & driver – whatever money be customarily charged by that class of men – which you

> know is not very great – likewise, of course, my Coach – hire or other necessary conveyance to and fro for the business in question. My staff-holder will be one of my pupils (now with me in London) who is a *good office-man* & for whom I should charge, if employed in the Office at any time, one Guinea per day, *to include all expenses*.[15]

He found his money was hard-earned.

> The Brierley-Wood (well named) has been a most awful obstacle to progress – indeed the last 38 Chains, from the Summit of the Hill, downwards, are quite inaccessible to the level & Staff, unless a regular road were cut through, which would occupy a great deal of time, besides placing the party to opening wood in a position to be blamed by the Duke's people – I never yet was in such a regular tangle of briars, binds and underwood, & I therefore *levelled round*.[16]

A fortnight later the unhappy man was in bed after an 'awkward though apparently ludicrous fall'. Poor Hamilton's catalogue of disasters continued. It began raining, and continued raining non-stop: 'The dykes are filling fast: every ditch becoming a Dyke, and every Dyke a River.' To add to his problems, the wind threatened to blow over surveyor and theodolite, while the roads were soon knee-deep in water. 'This country', he declared in despair, 'is only fit for ducks, Geese and human Amphibeae'. It is hard not to sympathize with Hamilton desperately scribbling his notes with a stumpy pencil in an increasingly soggy pocket-book. Other surveyors had even more dramatic stories to tell. The young Joseph Fowler came alarmingly close to ending his career as a young assistant. He was engaged in surveying the line between Lancaster and Ravenglass, across Morecambe Bay with its complex of river estuaries. The area he was surveying was cut by a channel no more than 150 yards wide at low tide, but at high tide the waters stretched out for nearly two miles. Fowler and his associates suddenly realized that they had been taking rather longer than they had intended over their work:

> The tide was then rolling in apace, and we had nearly a mile to run to get to the boat . . . However, off we set at a rattling pace; the first low place which had been covered by the tide was not more than a foot deep, and easily passed; then, tally ho! on for the next, which we knew was the deepest, and if we couldn't pass it return was impossible; however the first got over in about 18 inches of water, the next, who was at his heels, about 2 feet, the next deeper; I was the next with my level on my shoulder, and got clear of a tremendous wave, which, however, caught Marsh, who was at my heels, and nearly overset him.[17]

They all managed to return safe and secure to dry land, only to be endangered again when taking a coach across Morecambe Sands. A sudden storm blew up, and the first they knew of any trouble was when the coach came to a halt. Fowler got out to investigate and found the driver in tears and hopelessly lost. It was pure good fortune that they met a carrier who knew the Sands and was able not only to guide them to safety but also to provide them with the somewhat alarming news that when he met them they were cheerily driving straight out to sea!

No survey party, however, ever enjoyed – if that is the word – so dramatic a journey as the group that gathered in January 1889 to look over the proposed route of the West Highland line, across Rannoch Moor to Fort William.[18] There is probably no more desolate, woebegone stretch of country in the British Isles than this – a region of peat bogs, pallid grass tussocks and brown, brackish pools that offers no shelter whatsoever. It was across this uncompromisingly bleak landscape that a party of three engineers, the contractor Robert McAlpine, a solicitor and two local land agents set off on a bleak January day. The journey was to prove a catalogue of disasters. The first day was bad, but the next was worse. Plans had been laid and messages sent, but the plans were never realized and the messages seem to have got lost. They set off by coach which dropped them at Inverlair Lodge, where a guide was to meet them. No guide appeared, so they ventured forth on foot. Two of the company, having heard a rumour that the weather in the Scottish Highlands could be uncomfortable, took due precautions – they brought umbrellas. The next stage was by boat down Loch Treig, but by the time boat and boatman had been located dusk was settling over the hills. Off they went in a leaky boat, that had to be baled out using boots, for Lord Abinger's hunting lodge where a hot meal and comfortable beds were, they hoped, awaiting them. No preparations had been made – no fires, no beds, no meals. The next day brought near disaster.

They departed in the morning under lowering clouds for a 23-mile tramp across the moors. Long before nightfall exhaustion set in. John Bett, a local factor in his sixties, was the first to succumb and he was left in a makeshift tent, manufactured out of the two brollies. McAlpine decided to try to find help and was soon lost in the gloom of evening. Bullock, one of the engineers, also wandered away and he succeeded in finding a fence which he fell over, knocking himself out. Recovering, he had at least enough wits left to deduce that a fence indicated property and he followed it round to a shepherd's hut. The two shepherds set out and located the old man who by now was half-dead with cold. McAlpine meanwhile had marched steadily on and found his own way to civilization. The 'survey' was completed, but only just in time. The very next day a blizzard hit the region and enveloped the whole moor in snow.

It is always fascinating to dwell on the bizarre and the dramatic, but most surveys were, in truth, slow, plodding, mundane affairs. The greatest fear that most surveyors faced was not being trapped by the tide nor being engulfed by mountain blizzards, but the fear of not having plans ready in time to meet the parliamentary deadline of 30 November. A plan that was ready on 1 December was entirely useless until the next year. That was drama enough for most surveyors.

Any success at getting material ready on time was greeted with wild elation.

> I have just sent 4 Clerks to Thame to wait for tracings so that I shall have 8 there altogether tonight and 11 tomorrow with an able man to look after them so that I feel *comparatively* comfortable. White left Thame just in time to reach London at 4 *yesterday afternoon*!![19]

When it was found that it was not possible to complete the whole project by that dreaded deadline of 30 November, stern rebukes were duly issued.

> The solicitors have no doubt duly apprised you that they were unable to deposit the plans sections and other documents on the 30th November (yesterday) as required by the Standing orders of Parliament in consequence of the Surveyors having failed to furnish them according to their engagement.
>
> You were apprised from time to time of our fears that the Surveyors promises were not to be depended upon. If we had been authorised (even at a late period) to take the work out of their hands and proceed without them we have no hesitation in saying that we should have had everything required in readiness.

It certainly made for a busy life. John Fowler remembered those days.

> If I had to go seven miles to the works and back I walked, as the company could not afford a conveyance. If I had to get lunch at an inn, the company's limit was one shilling; and if the work was urgent, to finish drawings or specifications on a certain day, we constantly worked twelve hours a day.[20]

Although the work was hard, it did at least ensure that a good, enthusiastic young engineer need never be short of work.

> You have no idea how we were driven to get our plans ready for lodging at Preston before November 30. The hurry and anxiety were five times greater than with the Liverpool and Manchester line. For three nights none of us went to bed, and when all was finished every one was completely knocked up . . . The worst is that I can see no end to it, for the public estimation and enthusiasm for new railways and locomotive machines is duly augmenting; and I find that my opinion and service are in constant requisition.[21]

And it was not just the competent and qualified who benefited from the mad dash of surveying. *The Builder* complained of 'pseudo surveyors . . . utterly incompetent and yet are paid immense salaries . . . youngsters, hardly able to tell the right end of a theodolite, who are receiving two, three and four guineas a day and their expenses'.[22] They could, indeed, earn even more. One portly gentleman of no experience whatsoever boasted that he was offered five guineas a day, but it was in hilly country and he would have got out of breath, so he settled for three guineas on the flat instead.[23] But then almost anyone, it seemed, qualified for the job. A rough-looking Irishman applied to join the police. He was asked his job:

> 'Faith, and I am a railway surveyor!' 'A railway surveyor?' asked the clerk with some surprise. 'Yes, sure, your honour and didn't I always carry the poles!'[24]

Even then, if the survey was ready, the plans still had to be printed, deposited with interested parties and sent to Parliament. In the mania years, all was panic. Lithographers worked day and night, snatching what sleep they could on the tops of lockers or benches or on the floor. One company, in desperation, brought printers over from Belgium, but still failed to meet the deadline. Getting the plans deposited was just another headache, particularly if there was a rival company involved. Anecdotes of dirty dealings abound. One company tried to get its plans up to town on a rival's train, but various feeble excuses were used to refuse it. Matters were rapidly reaching crisis point, so the first company hired a hearse, black-plumed horses and an entourage of solemn-faced mourners. The coffin was loaded on the train and dispatched to London, complete with its load of plans and prospectuses. The Dudley, Neadeby and Trowbridge promoters thought they had got over the problem by hiring post-chaise and driver weeks in advance, only to find that when the great day came the carriage went at a snail's pace which no amount of cajoling or blustering would quicken. The driver's purse carried the opposition's shilling. The promoters leaped from the carriage, pulled the driver off his cab, gave him a sound beating and put him back in place. The rest of the journey continued at a good gallop.[25]

Inevitably some companies left everything to the very last minute. The office doors closed at 12 sharp on 30 November 1845, by which time 815 sets of plans had been deposited, but there were still optimists who hoped to make it 816.

> A post chaise with reeking horses, drove up in hot haste to the entrance. In a moment its occupants (three gentlemen) alighted, and rushed down the passage towards the office door, each bearing a plan of Brobdignagian dimensions. On reaching the door, and finding it closed, the countenances of all drooped; but one of them, more valorous than the rest, and prompted by the bystanders, gave a lond pull at the bell. It was answered by Inspector Otway, who informed the ringer that it was now too late, and that his plans could not be received. The agents did not wait for the conclusion of the unpleasant communication, but took advantage of the doors being opened and threw in their papers, which broke the passage-lamp in their fall. They were thrown back into the street. When the door was again opened, again went in the plans, only to meet a similar fate.[26]

It was not just plans that were needed by Parliament. The Act gave power to raise capital, so it was necessary to have an estimate of what the cost was likely to be. This involved a careful assessment of the earthworks – how much was to be dug out of cuttings, how much put up in embankments. Test bore-holes along the line of a projected tunnel gave at least a reasonable idea of the conditions likely to be met once work got under way, though as many an engineer was to find to his cost, they always seemed to miss the problem spot, the underground spring, the quicksand. Last, and certainly not least, the survey was also expected to yield an idea of how much cash would be needed for land purchase and compensation. The first estimate for the Great Western Railway shows the costs roughly broken down.

Act of Parliament	£50,000
Land	£320,000
Excavation	£625,000
Masonry, bridges, etc.	£300,000

Rails, chains, etc.	£538,875
Compensation	£50,000
Directors	£10,000
Tunnelling	£250,000[27]

Total estimated costs amounted to £2,143,875 and Parliament duly passed an Act authorizing £2½ million capital. If one looks at modern estimates for major projects, such as the Channel tunnel, prepared with the services of the most sophisticated computers, and sees how wildly actual costs differ from them, it is no surprise to find that matters were handled no better a century and a half ago. Final costs were to come out at £8,000,000. But there had to be an estimate, and once one starts looking at the problems that beset railway builders, perhaps even a 300 per cent error is forgivable.

CHAPTER FOUR

To Parliament

A new delay has occurred in the progress of our Bills – a new Crotchet has come across some Lord's head and a new Clause is sent to the Commons for approval – this utter wanton disregard of the interests of the parties waiting is really disgusting. The Tyranny exercised is as great as it could be under the most despotic Government and the only answer that one gets to the strongest appeals is insult and ridicule. If I had not been a Radical before, I should become one now.

I.K. Brunel, Report to the Great Western Railway, 25 June 1837

Having got its plans deposited, its estimates prepared and its promises of finance, however unreliable, at least entered in a ledger, the railway company had scarcely time to draw breath before being plunged into battle with its opponents. All kinds of varieties of opposition appeared. There were the rival transport systems – the turnpike road trusts and the canal companies – and increasingly, as lines spread across the country, there were rival railways. There were landowners, the nineteenth-century version of the twentieth-century Nimby – the 'Not in my back yard' opponent. There were all kinds of special interests to be appeased. And there were those who quite simply did not like railways. In the latter category came one of the most colourful characters of the age, Colonel Charles De Laet Waldo Sibthorp, MP for Lincoln.[1] He was what his friends might have called a 'Tory of the old school' and his enemies, or even disinterested observers, would describe as a high-handed, loud-mouthed, opinionated braggart. The locals seem to have regarded his often outrageous behaviour with tolerant amusement. 'Nobody cared what they did in those days. I can remember Colonel Sibthorp galloping through the market, upsetting the stalls and smashing the crockery. But he paid for it like a gentleman.' So that was all right, it seems. Even his obituary in *The Times* in 1855 could come up with nothing more flattering than a description of him as 'the embodiment of honest but unreasoning Tory prejudice'. His views could be neatly expressed in a single statement to Parliament: 'I have never travelled by railroad – I hate the very name of railroad – I hate it as I hate the devil.' He was not one to mince words. Railway schemes were 'dangerous, delusive, unsatisfactory, and, above all, unknown to the constitution of this country'.[2] It was useless for his opponents to point out that the conduct of railways was regulated by the Board of Trade – his views on the Board were much the same as those on railways. Happily for the railway interest, Sibthorp was regarded as a parliamentary buffoon whose opposition was not to be taken seriously. Other opponents were altogether more threatening.

Not surprisingly, the turnpike trusts who got their income from road-users, stage-coach owners, ostlers and a whole variety of other road interests greeted the imminent arrival of a railway with alarm. Their fears were not unjustified. On 22 July 1844, the day the railway opened between Bristol and Gloucester, the North Mail set out with the horses wearing black plumes and the coachmen in full funeral garb. It was never to run again.[3] In the early days, opponents made great play of the dangers of the newfangled machines. When the scheme for the London and Greenwich Railway was first mooted, the *Quarterly Review* of March 1825 published a much-quoted lampoon.

> It is certainly some consolation to those who are to be whirled at the rate of eighteen or twenty miles an hour, by the means of a high pressure engine, to be told that they are in no danger of being seasick while on shore; that they are not to be scalded to death, nor drowned by the bursting of the boiler; and that they need not fear being shot by the scattered fragments, or dashed to pieces by the flying off or the breaking of a wheel. But with all these assurances, we should as soon expect the people of Woolwich to suffer themselves to be fired off upon one of

> Congreve's ricochet rockets, as to trust themselves to the mercy of such a machine going at such a rate.[4]

A local horse-bus proprietor, George Shillibeer, set forward his case in a merry little jingle.

> These pleasure and comfort with safety combine,
> They will neither blow up nor explode like a mine;
> Those who ride on the railroad might half die with fear,
> You can come to no harm in the safe Shillibeer.[5]

But as more and more people travelled the railways without coming to any harm, such appeals faded away. Other arguments were brought forward. Investors in a turnpike trust declared that they had invested purely for the public good – now, with a railway in the offing, they would suffer a double blow. They would still have to pay for the upkeep of the road, but there would be no income from road-users. They were, they declared, ruined men.[6] Parliament was inclined to take the view that the investors in roads were, like the investors in railways, on the whole more interested in private profit than public virtue and that if their investment had turned out badly that was not the government's concern. There was, in any case, a good deal of evidence to show that in spite of protestations of ruin, the railways in creating a demand for travel were also creating a new demand for coaches on the road to get the passengers to the nearest station. Certainly, for a profession supposedly faced with imminent starvation, the coachmen were sometimes strangely reluctant to take such fares as were available. For many years, the picturesque little village of Steventon was the nearest point on the GWR to Oxford. There are still memories of those early days in The North Star Inn, with its signboard of an old broad-gauge locomotive, one of the places where passengers kicked their heels waiting for the coach to Oxford. It was often a long wait, as this letter to *The Times* in 1841 illustrates:

> A couple of minutes having been lost in seeking a porter, I crossed the rail to the coach, and found it already occupied by

> eleven outside and four inside passengers. The coach was, moreover, top heavy and unsafe from the quantity of luggage and merchandise which had been stowed upon it. On demanding a place, the coachman told me there was none for me, but on my being more peremptory he told me there was a second coach into which I got with two other passengers. In this coach we remained exactly one hour and a half. Having pressing business at Oxford, I got out to remonstrate with the coachman but he was not to be found. I asked for a chaise but a porter told me there was none. Entering the railway station, I remonstrated with the clerk, who expressed regret that the circumstances had not been speedily brought under his notice. I told him that it had occurred within his hearing and under his nose . . . He then conferred with another clerk and when I asked for the proferred remedy, his reply was 'You must wait for the arrival of the Bristol train which is due more than an hour and a half'. At this moment a superior clerk or inspector entered the office to whom I repeated my complaint. This person sent for the coachman and ordered him to drive off immediately for Oxford where I arrived at a quarter to ten instead of half past eight o'clock . . .[7]

A similar complaint of economic ruin about to descend on their hapless and innocent heads was made by the canal companies and their investors. There is a pleasant historical irony in reading the arguments of the Bridgewater Canal Company against the intrusion of the Liverpool and Manchester Railway into 'its' territory. Scarcely more than half a century earlier it had been the Duke of Bridgewater and his engineer James Brindley who had faced the wrath of the old river navigation companies at what they too had seen as a wholly unjustified intrusion into 'their' territory of a newfangled canal.[8] The arguments were surprisingly similar, and can be briefly stated: the new canal/railway could not possibly work and if it were allowed to go ahead would bring worthy public-spirited gentlemen to a pauper's grave. The fact that the two arguments were mutually exclusive did not worry the pamphleteers unduly. These were arguments for public consumption: the real battles would come later, waged by skilled barristers in parliamentary committee rooms.

Rivalry between different railway companies was equally fierce, and the arguments were again often couched in the same general terms. A railway company that had inveighed against a canal company and mocked its position as a self-seeking monopoly was quite happy to borrow these same arguments to defend its own monopoly in a region. Any stick could be used to beat the opponents. The Taff Vale Railway objected to the Vale of Neath's plans because it would take its trade away, with vessels taking cargo to Swansea instead of Cardiff, and then made great play of the 'dangerous' gradients on the new line. The Vale of Neath made short work of that argument.

> They took objections to the gradients, and not as they cared about the gradients, for the worse they were the better for them; but it was the only tangible, or speaking politically, their most positive ground of opposition.[9]

Inter-company rivalry sometimes led to one of the companies applying for and getting an Act not because it wanted to build a railway in that area but to keep competitors at bay. Requests for extensions of time for building then brought a quite new type of opposition – not from opponents, but from locals tired of procrastination who were actually desperate to see work start. In 1852, a petition was presented opposing the Great Western's extension of powers for land purchase for a branch to High Wycombe.[10] The locals pointed out that the GWR had been given powers in 1846, had them extended in 1849 and here it was three years later still asking for an extension. It was quite clear nothing was being done, and they quoted the chairman's own stated views as evidence.

> It may be extremely convenient to Towns and Villages to have Railways pass Close to them. If Railways are to be made for the purpose I have alluded to the onus of the Cost should fall not on the shareholders of the existing Railway Companies but on those who are to derive benefit and convenience from their construction.

In the end, the locals finished up doing just that – paying for the line themselves.

There were any number of local special interests who might object to a railway. The Provost of Eton College opposed the Great Western on the grounds that a railway would bring unsavoury moral influences to bear on the school, whilst the university authorities of Oxford and Cambridge insisted on special powers to reclaim errant undergraduates escaping from their books and scurrying off to the fleshpots and vice dens of London. Far more important to the railway companies, however, were the landowners along the route. The House of Commons still reflected the interests of the land-owning classes. The 1832 Reform Bill gave parliamentary representation to the newly developed industrial towns such as Manchester and Birmingham, but a Leeds Member of Parliament representing thousands of voters – not to mention a great many more non-voters – had no more power than the MP of some tiny rural community: the average size of a borough electorate remained under one thousand. And beyond the Commons lay the Lords, for any bill had to pass scrutiny by both Houses. Opposition by the gentry, and particularly the nobility, could sink any bill. The opposition to surveyors could reasonably be expected to reappear as opposition to the bill, so the promoters were realistically faced with two choices – avoid a troublesome landowner by a deviation, or buy him off. Contemporaries had no doubt about what was going on:

> Offers were made to, and accepted by, influential parties to withdraw their opposition to a bill which they had declared would ruin them, while the smaller and more numerous complainants were paid such prices as should actually buy off a series of long and tedious litigations.[11]

Even those local concerns who might be thought to benefit from the coming of the railway could not resist the temptation of an extra dip into the honey-pot. Brunel made his views on the opposition to the Bristol and Exeter Railway very clear.

> Exeter Corporation behaving in that unfair and illiberal manner which disgusted us so often in G.W.R. As we are sure of coming there and committed ourselves to a line they forget immense

> advantages and think only of how much they may hamper us and extract from us.[12]

Payment was not necessarily in cash. The Earl of Lichfield was rightly proud of his park and its splendid array of neo-classical follies. He was not averse to the railway as such, but he did insist that when the Trent Valley came to be built it should fit the grandeur of its surroundings. The stone bridge across Lichfield Drive has all the trappings: Ionic columns, balustrades, plinths with heraldic beasts and a coat of arms. The Earl had also insisted that the tunnel should be screened from view, but just in case it inadvertently caught the lordly eye, it too was suitably embellished, so that one end was a fair imitation of the entrance to a medieval castle and the other appeared in the form of an Egyptian temple.

The railways were prepared to go to such lengths when they were forced to do so, but only then. Whatever deals might be made in private, in public they played the stalwart businessman, counting every penny. Parliament got round to looking at the problems of paying compensation for land in 1845, and the solicitor to the Eastern Counties Railway proved to be of unflinching practicality. What, he was asked, about a railway running near to some beautiful old ruin, such as Fountains Abbey? What should be paid in compensation?

> It could be no injury to his pocket, and therefore he had nothing to receive.
>
> However hallowed the Ruin might be in his estimation?
>
> That is so. He must come to Parliament and pray of Parliament not to grant a Bill for a Railway to run so near so beloved a Place. But if it was a beautiful Spot, he should get a Station made near it, and turn it into Building Land.[13]

Given that attitude towards the beauty of the countryside and the historic treasures of the past, it was inevitable that railway promoters would fall foul of the Romantics. John Ruskin gave the Midland Railway a thorough tongue-lashing for its route down Monsal Dale.

> There was a rocky valley between Buxton and Bakewell, once upon a time, divine as the Vale of Tempe; you might have seen the Gods there morning and evening – Apollo and all the sweet Muses of the light – walking in fair procession on the lawns of it, and to and fro among the pinnacles of its crags. You cared neither for Gods nor grass, but for cash (which you did not know the way to get); you thought you would get it by what *The Times* calls 'Railroad Enterprise'. You Enterprised a Railroad through the valley – you blasted its rocks away, heaped thousands of tons of shale into its lovely stream. The valley is gone, and the Gods with it; and now, every fool in Buxton can be at Bakewell in half-an-hour; and every fool in Bakewell at Buxton; which you think a lucrative process of exchange – you Fools Everywhere.[14]

It might have consoled Ruskin to know that a century later the railway would be closed and the line would become a footpath. The Eastern Counties solicitor who showed so little concern for nature or history was loud in his protestations that the only course open to an honourable railway company was to pay a fair price for whatever land it needed, regardless of any other considerations. The tenant farmer should receive the same even-handed treatment as the Earl in his mansion. He must have known that such noble sentiments had little to do with railway practice. Another witness admitted to the committee that one landowner had been promised £5,000 – whether the line eventually came anywhere near his property or not – provided he dropped his opposition to the bill.

It made sense to the promoters to seek agreement – any agreement – with the rich and the powerful, though it might rouse the engineer, concerned only with the best possible route, to the heights of outraged indignation. Peter Lecount described payment to landowners as:

> Under all circumstances of fraud, delusion, and downright robbery, that can now be conceived. No means were left untried, no artifices unresorted to, and the most barefaced falsehoods unblushingly set forth in aid of one vast system of plunder.[15]

Not that the railways were averse to a little venality on their account – with a little help from their friends. In a dispute between the South Eastern Railway and the London, Chatham and North Kent over a proposed line, the Ordnance Factory at Woolwich gave its support to the latter. The factory's solicitor suggested ways in which that support could be made doubly sure.

> I think you should not overlook Captain Boldero (a member of this Board) in your allocation of the forthcoming shares; you can address them to him at this office, and mark it private.
>
> I spoke to Mr Bonham on the same point, but he feels some difficulty, as there is to be opposition in parliament to your lines.
>
> You can send some shares to him or not, as you please, and mark your letter 'private' and state that you do so at my suggestion.
>
> Whatever you intend for me you will also mark 'private'.[16]

Normally such affairs were rather more carefully hidden from view, but most contemporaries at all familiar with the railway scene were well aware, particularly during the mania years, how easily a few shares could be lost in the complex system of dealings that marked the age.

Eventually the time arrived when the various interested parties, complete with a regular phalanx of lawyers, presented themselves at the committee rooms of Westminster to state their cases. It was a cumbersome procedure. First a procedural committee would meet to make sure no one was breaking the rules; then, if all was well, affairs were passed to the select committees.

The select committee of the House of Commons would be made up of members who had a legitimate interest in the proceedings as their constituencies were affected and, to maintain a balance, 'disinterested' members. In the event, many of the 'disinterested' were to prove to be little more than pockets waiting to be filled. Both sides called witnesses and both sides came supported by banks of legal gentlemen. Having passed the Commons, the bill went to the Lords' committee, a rather more manageable body of five peers, none of whom had any direct interest in the outcome. Witnesses were called by both sides and were subject to rigorous cross-examination.

One of the first victims of the system was George Stephenson. When he had taken over the survey from William James on the Liverpool and Manchester line in 1824 he had been rushed for time and harassed by landowners and had inevitably to rely on assistants for much of the work, which he never personally checked. He might have felt a little dubious about some of the results, but there was nothing that could not be straightened out in a single conversation with a like-minded practical man of affairs. Stephenson was not, however, to meet a blunt, bluff, no-nonsense man like himself: he was about to face a remarkably shrewd, sharp-witted and wholly ruthless London lawyer, Edward Alderson, representing the bill's opponents. Alderson was going to tear poor old Stephenson's evidence to shreds.

Stephenson could have had little idea of what was in store for him. Other witnesses had been questioned closely but gently. Alderson clearly realized that only one man mattered: destroy George Stephenson's evidence and the rest would simply crumble away. It began badly, largely because Stephenson had confidence in his own knowledge of locomotives. He had been warned, however, that the public was not ready to accept such a ludicrous notion as that of a steam engine careering along at anything greater than human walking pace. The engineer tried to play the political game, but let out his own view that far higher speeds were possible. This enabled Alderson to raise the spectre of a monster unleashed on an innocent and unsuspecting public if the engineer was right. If the suggestion was wrong then, of course, it proved Stephenson's incompetence. But at least at this stage Stephenson was sure of his ground. When the time came for his questioning on the survey, he found the ground disappearing beneath his feet. He was quite simply unprepared to answer even the most basic questions.

What is the width of the Irwell here?
I cannot say exactly at present.
How many arches is your bridge to have?
It is not determined upon.
How could you make an estimate for it then?
I have given a sufficient sum for it.

By the end, it was all too clear how little of the survey work had been overseen by Stephenson himself, and how woefully he had been let down by his assistants. In his final address Alderson's invective reached full flow.

> Let us examine a little. Mr. Stephenson speaks of an arch at Lawton, which is to cost £375. How high is it to be? He does not know. At what rate per yard is it to be? He has not formed an opinion. Is this the evidence upon which a Committee of the House of Commons is to decide? If he will tell you the rate, we can estimate the dimensions. If he does not tell us either the one or the other, he is in this dilemma and there I will leave him. He is either ignorant or something else that I will not mention.

As the catalogue of errors was picked over item by item, the impression of total incompetence increased. It mattered little that Stephenson's plan to 'float' his railway over Chat Moss was, in the event, to prove practical – it sounded absurd as described by Alderson. Stephenson might have been given credit for knowing best, but his position was constantly undermined by the mistakes presented in the plans. For example, a bridge over the Irwell was to be built so low that even a modest flood would have all but submerged it. Alderson made the most of such errors.

> Did any ignorance ever arrive at such a pitch as this? Was there ever any ignorance exhibited like it? Is Mr. Stephenson to be the person upon whose faith this Committee is to pass this bill, involving property to the extent of 400,000 *l* or 500,000 *l*, when he is so ignorant of his profession, as to propose to build a bridge not sufficient to carry off the flood water of the river, or to permit any of the vessels to pass which of necessity must pass under it, and to leave his own Railroad liable to be several feet under water.

At one point Alderson declared himself 'astonished that any man standing in that box could make such a statement without shrinking into nothing'.[17] By that time, Stephenson would probably have been only too happy to oblige. Inevitably, the bill was lost, the survey had to be done all over again and a new bill brought back to Parliament. It was an example at the very beginning of the Railway Age that turning a bill into an Act of Parliament was no mere formality.

One of the problems of the early days was that no one was altogether sure what the locomotive could do or how it was to be operated. Here canals offered no guidelines. On the canals anyone could, in general, run a boat provided the appropriate tolls were paid. Some seemed to feel that the same system could apply on the railways. Captain Moorsom of the London and Birmingham was questioned on this point, and at first he tried to pretend that the notion was possible.

> Suppose an individual chose to put upon the London and Birmingham Railway, a locomotive engine, and he tendered to the company the toll of 2d per ton per mile, would the company demand from that individual any more than the 2d per mile for the passengers?
>
> . . . If a person were to come to me with that proposition, I should say, 'What more do you require?' If it is to run your engine from one end and to the other, it might be admitted on the terms stated of 2d per head per mile; if you must have water, if you must have coke and turn-plates and stations, and so on, then it would be different.

Eventually, however, he gave the straight answer.

> Would it be possible for any individual to work a locomotive engine on the London and Birmingham Railway without the use of these conveniences? No.[18]

What is curious is to see how many questions on aspects of running the railway the committee managed to introduce, when all the company wanted was the power to raise extra funds. For example, it managed to slip in a clause which specified a 40-shilling fine for anyone smoking 'in and upon the carriages' and in the station. Later enquiries, however, tended to centre on much simpler questions. Once railways were well established and running methods clearly indicated, arguments came down to basic disagreements between interested parties. Engineers expected a grilling and usually got it, but they were well prepared and soon the parliamentary committees were proving a very valuable source of income for young engineers. When John Fowler, still in his twenties, set up as an independent engineer in London in 1844 he found he spent more time giving evidence to committees than he did in actual engineering. In 1847 he appeared as an expert witness in thirty-eight cases, and he freely admitted that he was as happy to attack a case as to defend it. In later life, he was to impress on his pupils the necessity of arguing a case with brevity and clarity. He was also, it seems, a sensible young man for when a company offered him £20,000 to prepare a survey of a major route to be ready in a matter of weeks to present to Parliament, he declined the offer.[19]

Appearing before committees was an accepted part of the lives of engineers, and they came to regard the probing and sniping of barristers as an unpleasant necessity, rather like surveying in the rain. It must have come as rather more of a surprise to the local gentleman who came up to London to lend his twopennyworth of support to the proposed local railway. He was there, he said, simply to represent the view of the ordinary people of the area who rather liked the idea of having a railway. He had hardly had time to make this seemingly innocent statement before the attack began.

> Do you know Mr Taylor? – No, I do not know him.
>
> I thought you spoke to the feeling of the district round Alcester.
>
> If I tell you there is a gentleman that lives between Alcester and Warwick not in favour of this line, you tell me that you do not know him; do you know Mr Greenhill? – No, I do not know Mr Greenhill.
>
> Or Mr Cowley? – I do not know them as owners, they are certainly not owners.
>
> These are gentlemen occupying farms on the line; do you know Mr Parks? – No.
>
> Then really you are not able to say what the feeling of the district is? – Yes, but I do not know these gentlemen.

You are able to say so, only subject to your not knowing these gentlemen who represent a large part of the line.[20]

The early railway bills stirred up general interest – they were new and exciting. Citizens of London could read of locomotives rushing across the land at great speed, but few of them had ever seen one. The novelty, however, wore off soon enough, and as the plans fluttered down on Parliament like leaves in an autumn gale, it must have seemed that every committee room in both Houses was fully booked up with railway business. But what appeared to be just another tedious argument over a purely local line could still be a matter of intense interest to the locals themselves. In 1861 two competing bills – the Newcastle, Derwent and Weardale and the Blaydon and Conside – came before Parliament and dominated the news pages of the local press.[21]

First there were the arguments for the different routes. The Derwent Valley route would cut the journey time to Liverpool to six hours, serve industry and enable merchants and businessmen of Newcastle 'to get ready access to pleasant country seats'. There were denunciations of the rival, which was under the patronage of the North Eastern, who made every effort to ensure goods used its tracks. 'The Newcastle and Carlisle receive black mail from the North Eastern, for closing their own line to goods . . . so as to screw the greatest possible mileage out of the public.' Then there were allegations of foul play in the petitions – 'signed four times by one gentleman, and twice over by three or four others'. The gloves were off from the start and local interest followed the case down to Parliament: 'Committee Room No. 5 and the lobby, resemble the Newcastle Quay, so many well-known faces are to be seen there.' In the end all arguments came down to one simple proposition: could a new railway come in and break up an old monopoly? Parliament decided it could – the Newcastle, Derwent and Weardale won the day – but by then the arguments had taken up so much time, nearly a month in committee, it had become known as the 'Wearydale' case.

Yet even this railway marathon seems like a sprint compared with the really great contests, and of all the inter-company arguments that occupied Parliament in the nineteenth century, none raged more fiercely, nor more expensively, than that over the Great Northern Railway. It was an ambitious scheme, a railway equivalent of the Great North Road – a gargantuan scheme which involved authorization of 327 miles of track, of which 186 miles consisted of the principal main line from Yorkshire to London. With such a vast area to cover, and so many properties involved, the company must have expected a degree of opposition, but it was the threat to existing railways that was to create the real problem. Eastern Counties was concerned because of the lines from the east coast ports of Hull and Grimsby; even more importantly the new route carved straight through the heart of George Hudson's kingdom. Worse still it held out the promise of a better, faster service to York – the Great Northern's line through Doncaster would be 30 miles shorter than the route through Derby. The rivals fought a massive parliamentary battle and it is now almost impossible to get an accurate assessment of the costs; but it seems certain that between the depositing of the plans in November 1845 and the passing of the Act in the following June, each side spent around half a million pounds. At least the Great Northern had an Act of Parliament to show for its endeavours; the rival had only a strong sense of resentment, which the passing years did little to alleviate. There was a price war, which was at least good news for the public, with the Great Northern promising to undercut all rivals and on one famous occasion, when running powers into Nottingham were in dispute, a Great Northern train that ventured into that station found a Midland locomotive hastily run in behind it. The Great Northern train remained imprisoned for seven months, while the companies spent yet more money on the wrangling of lawyers.

Getting a bill through Parliament was time-consuming and money-consuming. A lot of people, however, did very well out of it. Lawyers thrived and unscrupulous MPs could – and did – take bribes. By 1845, no less than 157 members sat on railway boards and were handsomely paid, not in many cases for their value to the running of a railway but for their influence in Parliament. The committees of Lords and Commons no doubt did valuable work in weeding out some of the more preposterous schemes and sending them packing, but their prime purpose was to see if a line had enough cash and expertise to ensure completion and to make sure that certain special interests were looked after. There was very little concern about the overall pattern of railway development in the country, no sense of fitting a line into a large plan. But for the individual companies, the day when the bill received the Royal Assent was a day of great celebrations. The preliminaries were over: the main action could now begin.

CHAPTER FIVE

Founding Fathers

The Stephensons were inventive, practical, and sagacious; the Brunels ingenious, imaginative, and daring. The former were as thoroughly English in their characteristics as the latter were perhaps as thoroughly French. The fathers and sons were alike successful in their works, though not in the same degree. Measured by practical and profitable results, the Stephensons were unquestionably the safer men to follow.

Samuel Smiles, *The Lives of George and Robert Stephenson*, 1874 edition

A railway is a complex institution, far too complicated for any one man to be lauded as its author and sole progenitor, but if there has to be a name attached then it must be that of the chief engineer. He took the ultimate responsibility for the decisions that were to give a line its individuality and character. He decided the line it was to follow, fixed an embankment here or a cutting there, decreed that the tracks should soar over a valley on a grand viaduct, plunge through the hills in a long, dark tunnel or simply bypass all obstacles in great sweeping diversions. He advised on rails and sleepers, the siting of stations and the ordering of locomotives. If any personality was to be stamped on a railway then it was that of the chief engineer - and no two men displayed more powerful personalities nor left more distinctive marks than those two great pioneers of the early years, George Stephenson and Isambard Kingdom Brunel. Samuel Smiles, whose thumbnail sketch opens the chapter, was, not surprisingly, a Stephenson man. Stephenson epitomized the virtues set out by Smiles in *Self-Help*. Here was a man who had raised himself to eminence entirely by his own efforts, a man without formal education who had proved to the world that the practical man of sound common sense could achieve anything that the cultured and refined could manage - and a great deal more. He was, for Smiles, the epitome of nineteenth-century man. Certainly, even now, looking back at George Stephenson and his career, it does appear clear that he was not just a man of his time: he was a man lucky enough to be in the right place at the right time. Stephenson seems to have been created for glory in the field of engineering. Brunel is quite different. That he was at least as well suited to a career in engineering as Stephenson is obvious from his accomplishments. But equally one can imagine Brunel, given different circumstances, shining in other ways - as a great scientist perhaps or a successful surgeon, in anything which required a combination of brilliant imagination and practical skills. They were not the only engineers at work in the early days of the railway, but they were the giants who dominated the scene. They represented two extreme positions, two very different approaches to railway building. In retrospect, one might feel that neither was wholly right nor wholly wrong, but between them they covered the full range of possibilities. No history of railway development could be complete without at least a brief glance at their lives and work.

George Stephenson was born in 1781 at Wylam, a colliery village on the north bank of the Tyne.[1] He was brought up in a world where natural abilities could easily have been lost from view in the hard grind of the miner's life, for it was a virtual certainty that George would one day follow his father to the pit. However, he was to arrive there not as a grubby lad crawling through the dark places of the earth, but as a servant of the mechanical marvel of the day. The eighteenth century was the first great age of the steam engine: the elementary ponderous beasts had begun pumping water from the deep pits early in the century, but by the 1780s James Watt's improvements had greatly increased not only their efficiency but also the number of uses to which they could be put. Stephenson's father worked on just such an engine. And every day of his young life George looked out at the Wylam wagonway, where horses pulled the coal wagons literally past his front door. Steam and railways were a part of his life from infancy.

He was always spoken of as a big, strong boy, though one harbours the suspicion that in some of the stories memories of his strength grew in direct proportion to his own growing fame. He followed his father to the colliery, not as an underground worker but as one of the men who helped to keep the steam engines running. By 1798 he was in charge of a pumping engine, and by 1801 he was a 'bankman' who had the responsible job not only of lifting trucks and coal out of the mine, but also of controlling his engine to ensure a safe journey for the colliers going to and from work. He was paid a pound a week, made extra as a part-time cobbler and considered himself well enough off to be able to marry. That was in 1802. The following year his son Robert was born and he himself was given a new and, in retrospect, important job. The tramway trucks had for some time been rolling down the hill with their loads of coal for the waiting ships on the Tyne. These coastal colliers had, as a rule, a one-way trade - there were no coals to be brought to Newcastle, and seldom very much of anything else. So they were forced to carry ballast which had to be unloaded when they arrived in the north-east to take on coal. A steam engine was installed at the top of the hill to haul the ballast up the incline and the young man was put in charge. George Stephenson, rails and steam had come together.

Stephenson's progress for the next few years was steady rather than spectacular. He was a diligent worker and a self-educator, improving his basic skills in the three Rs. Perhaps nothing seems more surprising to the present age than the notion of a great engineer whose formal education scarcely went beyond that of a child leaving primary school. He never regretted the lack of further education himself, and would certainly have agreed with the views expressed by one of his great contemporaries, John Rennie, even if he disagreed with him on almost everything else.

> My father wisely determined that I should go through all the gradations, both practical and theoretical, which could not be

> done if I went to the University, as the practical parts, which he considered most important, must be abandoned; for he said, after a young man has been three or four years at the University of Oxford or Cambridge, he cannot, without much difficulty, turn himself to the practical part of civil engineering.[2]

One of Stephenson's characteristics was a certain stubborn obstinacy, particularly when faced with what he scornfully referred to as 'London men'. His world had little time for theories – his approach was essentially practical. The antagonism was often mutual. The young engineer Charles Vignoles, who came to engineering via Sandhurst and the army, gave his view to another young engineer, Josias Jessop, who had also quarrelled with Stephenson:

> I also acknowledge having on many occasions differed with him (and that in common with almost all other engineers) because it appeared to me he did not look on the concern with a liberal and expanded view, but with a microscopic eye: magnifying details, and pursuing a petty system of parsimony, very proper in a private colliery, or in a small undertaking, but wholly inapplicable to this national work.
>
> I also plead guilty to having neglected to court Mr S's favour by crying down all other engineers, especially those in London, for, though I highly respect his great natural talents, I could not shut my eyes to certain deficiencies.[3]

Vignoles could hardly be described as an impartial observer, since he had been sacked by Stephenson, allegedly for taking inaccurate levels during work on the Liverpool and Manchester Railway, but more probably because of the very different attitudes that come across in the letter. In many ways, this was the least appealing side of Stephenson's character, but it had its origins in his early years. His knowledge was slowly and painfully gained, and if he came to fortune later in life that has to be balanced against the many years of working as a colliery engine-man. He was proud of the way he had come to knowledge and understanding by his own unaided efforts, but was sufficiently aware of its drawbacks to devote much money to his son's education. He had another reason to concentrate all his efforts on young Robert. In 1804, his baby daughter died to be followed, in the next year, by his wife Fanny.

George Stephenson's reputation as a good practical engineer spread around the small world of the Tyneside collieries. He was the man to put an engine to rights – and in time he received his reward with appointment as engine-wright at Killingworth with overall responsibility for the machinery in the extensive series of pits of the Grand Allies. It was in this new role that he was able to visit the recently opened Middleton Colliery Railway in Leeds, and to call in to see the operations of engineers such as William Hedley of Wylam who were starting to work on steam locomotives. Trevithick had sent a locomotive to the north-east but it had not proved a success. The Wylam engines, *Wylam Dilly* and *Puffing Billy*, were altogether more promising. If the manager of Wylam could build locomotives, so too could the engine-wright of Killingworth. Stephenson began work on his first engine in the autumn of 1813.

Over the years there has been controversy over the title 'The Father of Railways'. The arguments are nonsensical. One can point to the pioneers of the tramway age, to Trevithick and many others, but railways did not spring unaided from a single brain. Sadly the enthusiasts, as well as championing their favourites, tended to belittle the opposition. Stephenson, they said, had done no more than borrow Hedley's ideas. In fact, the first Stephenson engine, *Blücher*, owed a good deal more to the Blenkinsop and Murray engine of Middleton than it did to Hedley, though Stephenson abandoned the rack and pinion. Stephenson enthusiasts have been equally keen to attribute innovations to their hero, such as steam blast, where steam exhausting from the cylinders escapes via the chimney, thus increasing the draught to the fire. The idea had, in fact, already been tried by Trevithick but proved ineffective on its own in improving efficiency. Stephenson tried it, according to a contemporary, and found the effort dramatic but unhelpful. 'The steam thrown in this manner into the chimney acts as a trumpet, and certainly makes a very disagreeable noise'.[4]

In looking at the slow changes of these years what is so striking is the way in which progress was made. Design was by trial and error and experience. There were few, if any, text-books, and even if there had been, the men who were doing the work would often have been unable to read them. Travel was slow, difficult and expensive – hence the need for railways in the first place. A trip to Leeds for example, to view the pioneering Middleton Colliery Railway, was quite an excursion, even for men like George Stephenson who lived in the north-east, so few bothered to look for ideas outside their own immediate environment. The fact that, in France, men such as Marc Seguin and Henri Gifford were producing important innovations in locomotive design was quite unknown to the men of Tyneside – for all they knew of it, they might as well have been working on the far side of the moon. The empirical approach had its advantages: men like Stephenson were very unlikely to be found pursuing chimeras. But it could also lead to a narrow parochialism. Where a theoretician such as Brunel might begin by asking 'What is the best possible gauge for a modern railway?' and then reach the conclusion that a broad gauge offered many advantages, such a question probably never entered Stephenson's mind. His first engine was built to fit the rails already in place, and as new railways were added so the old system was simply extended. The odds on anyone starting from scratch and arriving at a figure of 4 ft. 8½ in. – or, if a Frenchman, 143.5 cm. – are remote. Yet when it came to the 'battle of the gauges' it was the 'illogical' measure that won the day – on the logical grounds that there was more track already laid to that measure than there was to the other. Here one can see the strengths and weaknesses of two great men. Stephenson was an unquestioning follower of tradition, with a keenly developed sense of what was practical in the real world. Brunel was unashamedly an innovator, with theory playing a significant part, prepared to cast the muddied waters of tradition to the winds and losing its benefits at the same time. Inevitably this is a generalization to which there are many exceptions.

One of the most revealing incidents in George Stephenson's career had no direct connection with steam engines or railways, but undoubtedly had its effect on the engineer's attitudes. In 1812 an appalling explosion devastated the Felling Colliery and ninety-two men and boys died. It was caused by methane gas and a naked flame: the hunt was on for a device that could safely light the miner's way. A committee approached the scientist Sir Humphry Davy who came up with his famous safety lamp. At the same time Stephenson began work on a lamp of his own, and tested it in the only way he knew. Lamp in hand he marched up to 'a blower', a source of escaping gas. The light remained steady, there was no explosion. Davy was awarded 2,000 guineas for his trouble, Stephenson 100 guineas, but the men of the north-east felt the injustice of this and raised 1,000 guineas for their man. Davy was furious and denounced Stephenson as an imposter, on the grounds that it was self-evident that a semi-literate collier could not possibly appear as the equal of a genuine man of science. The experienced mine owners and managers, well versed in the practicalities of the affair, stood firm. For Stephenson it was grim evidence that while he could rely on the support of his own kind, the experts – especially London experts – were liable to prove implacable enemies.

Meanwhile Stephenson continued working on into the 1820s, developing both the steam locomotive and the railway system on which it ran. Between 1814 and 1826 he was the only engineer in Britain building locomotives.[5] In collaboration with William Losh of the Walker ironworks he devised a new type of cast-iron rail which proved to be a great improvement over the other varieties. Despite his involvement and financial interest in this new rail, as soon as he was introduced to the wrought-iron rail he saw its advantages and recommended it instead of his own – with a considerable loss of income. Some other vested interests were not so adaptable.

> With respect to the Stockton and Darlington Railway Company advertising for *cast-iron rails*, it was merely to please a few of the subscribers, who have been brought over by some of the *cast-iron founders*, but they have only advertised for one-third to be cast-iron.[6]

Given Stephenson's increasing concern in the development of railways and locomotives, the Stockton and Darlington no longer seems like some sudden, inspired happening, but takes its rightful place as part of a continuous development in both Stephenson's career and the history of railways. It was one step – but a huge one. There was to be Stephenson's stumble over the ill-fated Liverpool and Manchester survey, but by 1825 he was generally accepted as the foremost engineer of the day, a man of irreproachable honesty. The one dark stain on his subsequent career was the way in which he was manipulated by the unscrupulous George Hudson. Stephenson would arrive at a meeting of shareholders and speak openly and frankly of what he saw as the engineering virtues of the line on which he was engaged, and Hudson would see that the assumption was made that the financial virtues of the line would prove equally sound.

In his later years George Stephenson moved to Leicestershire and also became a colliery owner. Later he returned to a farm near Chesterfield, where his greatest triumph was the successful growing of a straight cucumber. He died in 1848, rich in money and richer still in honour. He was not a great innovator but he was a wonderful stayer: he kept locomotive development moving forward when interest in the engines had all but collapsed, he was a tireless promoter of the idea of railways and he was a man who, when faced with a seemingly impossible problem, worried away at it until he had found a solution. The best-known example was the crossing of Chat Moss by the Liverpool and Manchester. Alderson had, as we have seen, a particularly fine time ridiculing the plans for 'floating' the railway across the morass.

> It is ignorance almost inconceivable. It is perfect madness in a person called upon to speak on a scientific subject to propose such a plan.[7]

Given the mauling he had received, Stephenson could have been forgiven for deciding to give Chat Moss a wide berth. But whether it was conviction in the rightness of his own plans, simple stubbornness or a natural desire to prove the experts wrong, he tackled the problem head on. And it was a problem: 12 square miles of peat bog that locals solemnly declared was bottomless. John Dixon, one of Stephenson's assistants, demonstrated that theory when he slipped off one of the plank walkways and began disappearing into the mire, before being dragged to safety. The crossing of the bog was to prove the value of persistence.

The bog was higher in the centre than at the edges which presented a problem when it came to digging drainage ditches. The deeper the ditch had to be cut, the greater was the tendency for the black ooze to rise and fill it. Stephenson's solution was to order sets of barrels from Liverpool. The ends were knocked out and the barrels nailed together to create a crude but effective drainage pipe beside the proposed route of the railway. The great problem remained, however, of how to lay tracks across the morass. Here the engineer conceived the notion of floating his railway on a gigantic raft of hurdles, heather and moss. When the raft was stabilized, gravel was laid on top to take the rails. At the Manchester end, however, there were problems. A bank would be raised as high as 3 or 4 ft. only for it to sink back into the bog. It seemed that nothing would stabilize the bank and the members of the board of the infant railway company began to despair and even considered abandoning the entire project. But Stephenson carried on, organizing his work force to cut dry turf and pile it on until at last the turf stood proud of the Moss – and stayed there. When the bank was finished it was estimated that 670,000 cubic yards of material had been tipped on to Chat Moss of which less than half appeared above the surface. If the committee rooms in Westminster saw Stephenson at his nadir, this bleak, dark, seemingly featureless expanse of flat bog saw him enjoy his triumph.

George Stephenson's place as one of the great pioneers has never been seriously challenged, but his son Robert was to earn his own place in the pantheon. Like his father he was a skilled mechanical as well as civil engineer. The *Rocket* alone would have ensured him an honoured place in the railway Hall of Fame, for it was the locomotive which set the standard for future developments. But he was to do very much more and was to produce some of the most remarkable monuments of the Railway Age, notably the bridges at Conwy and across the Menai Straits. In achievement he was his father's son: in personality and in approach he was very much his own man.

Robert Stephenson received a good education, first at the local village school, then at a private school in Newcastle which involved the young boy in a daily journey of 20 miles by donkey. Robert helped with the education of his own father, bringing home and reading from technical treatises, while George in his turn gave the boy the practical education that had no place in the middle-class schooling of the day. Robert, as he grew up, alternated study with hard work – six months at Edinburgh University was sandwiched between expeditions surveying with his father on the Stockton and Darlington and with William James on the Liverpool and Manchester Railways. He must, however, have felt the need to be his own man and when the chance came to work at re-opening silver and gold mines in Mexico he was full of enthusiasm. Bearing in mind that the alternative was helping to run the locomotive works that bore his father's name, the urge to get away must have been real indeed. The Mexican adventure was not a great success, but it had an odd outcome. In Cartagena, while waiting for the ship home, he met an English engineer down on his luck and trying to raise the fare home. It was Richard Trevithick. The older man's career was heading steadily downward; the young man was on the way up. It was a poignant, and no doubt uncomfortable, moment.

On his return, Robert was soon absorbed in one of the most ambitious of projects. Still in his twenties, he was appointed chief engineer for the main line from London to Birmingham. It was a line that was to be built to stand the test of time – it was to be kept remarkably level throughout its length by means of embankments and cuttings, tunnels and viaducts. Where a bend was needed, it was always built on a generous curve, rarely less than a mile in radius, and there was a cant on the corners, with the outside rail slightly higher than the inner. This was scarcely necessary for the tiny locomotives and short trains of the age, but surprisingly it has proved adequate for the high-speed expresses of the twentieth century which now travel the line. In many ways it was a truly forward-looking railway, yet Stephenson clung to some of the old ways. He still proposed using the stone blocks of the tramway age, where stone was available, instead of wooden sleepers, which is surprising seen from our perspective. He began in high optimism. When giving evidence to Parliament he was asked whether he anticipated any problems with the Kilsby tunnel, which was to be nearly a mile and a half long. He replied that it would be 'Very easy indeed: in all clays it is very easy to tunnel, unless they be a great deal mixed with sand.'[8] When work actually began he was much more cautious, and when it was finished he probably never wanted to hear the word 'sand' again.

Work on the tunnel began in June 1835 with the sinking of the two great ventilation shafts, each 60 ft. across.[9] In October Stephenson dropped his first hint of quicksand as it had been met with in an earlier canal construction. This was not the quicksand of popular fiction which sucks victims to their doom, but an area where sand meets an underground spring, with the consequence that as soon as an area is excavated the sand flows back in again. The solution is to pump the area dry and wall it off. By December he had arranged for a 'Quicksand Engine' to be set to work and was generally optimistic. By February things were less happy: in one place 'the water came in on us in such quantities that the men could no longer remain in the pit' and 'the sand broke in on us again'. To add to his problems, the tunnel contractor Joseph Nowell died and his sons were unwilling to carry on without him: Stephenson had to take on the work himself. By now the area was called 'Quicksand Hill' and more and more shafts were being sunk down to the sands, and plans were being laid to set seven pumping engines at work. It was reminiscent of his father's problems at Chat Moss. However, pumping would succeed, given time and enough pumps. Gradually the water levels dropped, but it was not until January 1837 that he could announce that 'The drainage of the quicksand is now completed.' The first 588 yards had taken eighteen months – the next 1,835 yards took a mere six uneventful months.

A quite different aspect of Robert Stephenson's character and skills appeared in another great project, the line from Chester to Holyhead. This involved the construction of two bridges – one across the Conwy River, the other over the Menai Straits.[10] Here one finds a wholly new approach, quite unlike what one might have expected from the elder Stephenson. The problems were immense, and first approaches were, to say the least, timid. Thomas Telford had already spanned river and straits with two impressive suspension bridges, and serious consideration was given to the notion of using the Menai bridge. This had two lanes, and the idea was to keep one for road traffic and convert the other to rail. It was obvious from the first, however, that it was not suitable for locomotives. So a train would have to stop at one side and then the coaches or trucks would be unlinked, pulled across the bridge by horses and then attached to a second locomotive on the other side. This wildly impractical scheme was soon cast aside, and Stephenson turned his attention to alternative crossings at Menai. The seaways had to be kept open, so to avoid one great span, Stephenson began to investigate a route which could use the Britannia Rock, conveniently situated in the middle of the channel between Anglesey and the mainland. This would take a central pier, and Stephenson showed from the first that he was willing to take advice from friends and colleagues. Many commentators have tried to invent a bitter feud between Brunel and the Stephensons. Certainly they had serious differences and fought as hard as they could to get their own views across and their own lines on to disputed territory, but there was real friendship and respect between the two men. Brunel made a number of helpful suggestions and travelled up to Wales to lend moral support during the final crucial stages of construction. Most ideas failed to meet the Admiralty's requirement that nothing should get in the way of the biggest sailing ships. The answer, when it finally came, was wholly original. In effect the bridge was designed with rectangular tubes, like great girders, but with the trains running inside them. What distinguished the project was the preliminary work and the use of advanced technology.

Stephenson received enthusiastic support from the ship-builder William Fairbairn. He invited Stephenson to see an iron ship being made and offered to display the remarkable rigidity of iron by propping up a 220-ft.-long ship at bow and stern, leaving the centre unsupported. Stephenson and Fairburn agreed that a tube bridge was possible, but it was necessary to decide what shape the tube should be and how it should be constructed. Stephenson decided to answer these questions by using a modern method: laboratory experiments with models of different sections. Stephenson perhaps remembered the parliamentary mockery his father had faced when proposing his novel floating railway and he was determined to avoid this. He sought to allay the scepticism which often met his ideas:

> I hope the Committee will not consider the idea as chimerical because it is new; it is only, as Mr Randel stated, substituting an iron tube for the purpose of getting a rigid platform.

In the end he got his way, but had to add wholly superfluous suspension chains to the design. The ports through which the chains were to pass can still be seen at the tops of the towers. The construction technique was truly remarkable. The tubes were riveted together *in situ*, then floated into place and gradually raised using hydraulic jacks. The Menai bridge was a massive enterprise – 1,800 ft. long and standing over 100 ft. above the water – so the first efforts were made at Conwy, a more modest 425 ft. long and much lower over the water. Brunel arrived on the scene to give his advice and the first of the great tubes was floated out at high tide on 6 March 1847.

> The tube, being lifted by the pontoons, began to move off, snapping the small ropes that kept it back; it glided quietly and majestically across the water in about twenty minutes . . . At eleven o'clock the deep and rapid Conway was an impassable gulf, and in less than half an hour it was spanned by an iron bridge.[11]

The work was not trouble-free, but it was in the end an undoubted triumph both for Stephenson personally and for Victorian engineering.

In many ways, Stephenson was closer in outlook to Isambard Kingdom Brunel, who was so free with help and advice throughout the project, than he was to his eminent father. Brunel is the best known of all the great engineers of the nineteenth century, though his fame does not rest solely on his railway work.[12] George Stephenson and Brunel are accepted giants of the age but they could scarcely have been more different in character, temperament, ideas and approach to their work. The one trait that they had in common was an overwhelming self-confidence: this could lead them to a pig-headed obstinacy in following their own inclinations, regardless of advice and argument. But without it, could they have pushed forward their own, often revolutionary ideas? Brunel has been widely criticized for two decisions that were to prove expensive mistakes – the decision to use the broad gauge for the Great Western Railway, with rails set 7 ft. apart, and his decision to try out the atmospheric railway on the South Devon line.

To get a perspective on the controversies that surrounded Brunel throughout his working life, one has to look back to his early years and his upbringing. His father Marc Brunel had a breadth of interests that almost rivalled his son's. He came to Britain as a refugee from the French Revolution with a plan for making blocks – essential parts of a sailing ship's rigging – in standard sizes by mass production. The plant was established at Portsmouth in 1808 and remained in use for 145 years – the building still stands in the dockyard, with a few isolated machines *in situ*. Marc proved able to turn his hand to civil as well as mechanical engineering projects, and was responsible for the first tunnel under the Thames – a project which almost ended in disaster, and which nearly cost young Isambard his life. The young Brunel was brought up in an environment in which innovation was all but taken for granted. Unlike his contemporaries, however, he was given a formal education designed to fit him for his future career. He received the standard education of any English boy of the middle classes, but at the age of 14 he was sent to France to receive a technical education then virtually unobtainable in England. He failed to enter the prestigious Ecole Polytechnique, but he did receive a sound grounding in both mathematics and theoretical as well as practical engineering. This is not the place to describe his early years working alongside his father but clearly they had a considerable influence on him. His father was a great engineer in his own right, and above all taught him the virtues of perseverance and self-reliance.

Brunel's achievements are too well known to require setting out in detail here, but it is instructive to look at his 'mistakes' and his working methods. No one can seriously agree, with hindsight, that the decision to opt for the broad gauge was right – it cost a great deal of money to lay down, and in the end had to be ripped up. But how did things look when the decisions were being taken? The Stephensons were busy spreading their 4 ft. 8½ in. gauge down from the north-east, but it was still not certain that a coherent rail system of interlocking lines would ever emerge, though the canals had clearly shown the disadvantage of not having just such a system, with different sized locks on different waterways proving a barrier to successful trading. The canals, however, also provided another historical lesson. Many of the waterways had been built using the narrow locks based on those of the pioneering engineer James Brindley. Acceptable in the eighteenth century, they were to prove inadequate in the nineteenth. More than one engineer felt that in blindly following a system developed for the limited world of colliery trucks and horse-drawn vehicles, the Stephensons were similarly laying up trouble for the future. Brunel came to the Great Western with what he saw as a clean slate, the only constraint being the possibility of reaching agreement with the London and Birmingham to share their Euston terminus. He felt that this could be overcome by each company laying a third rail.

> I do not see any great difficulty in doing this but, undoubtedly the London and Birmingham Railway Company may object to it and in that case I see no remedy – the plan must be abandoned.[13]

At this stage, Brunel was ready to give up his broad-gauge ideas for the great cause of unity, but when it became clear that the two companies were not going to agree, then there appeared to be a logic in a system of distinct areas providing self-contained systems. The south-west would be broad gauge; north of London would go to the Stephenson gauge; and already East Anglia, in the shape of the Eastern Counties, was going its own way. Brunel can be accused of lack of vision in not foreseeing a unified rail system for the whole of Britain, but in many ways there was sense in his notion of establishing at least one region which would have what he regarded, with some justification, as the best system – one decided upon on grounds other than a simple adherence to an old tradition. Although Brunel was no locomotive engineer – his early specifications for locomotives gave their designers almost impossible tasks – the broad gauge offered great advantages. Daniel Gooch, who at the remarkably young age of 21 was put in charge of the locomotive department of the GWR, had already experienced some of these advantages before he joined the company, when he had worked on a design for a Russian locomotive. Russia had yet another variation of gauge – 6 ft. in this case – and Gooch was able to keep the whole of the firebox and the motion within a 6 ft. space. Gooch's *Firefly* class broad-gauge engine took part in a trial against the best of the Stephenson standard-gauge engines in 1845 in front of the commissioners trying to decide for Parliament between the rival gauges. The broad gauge emerged triumphant. Had Brunel been first into the railway field his gauge – now extended by a quarter of an inch to 7 ft. ¼ in. – would have produced a fine railway offering greater speed, better comfort for passengers and more efficient freight movement. The tragedy from Brunel's point of view was that it was not the first. He had asked the wrong question. He should have enquired not if his version was the best but whether it was too late to change.

Similarly Brunel's decision to mount the rails on longitudinal sleepers is now seen as wrong, but at the same time Robert Stephenson was still using stone sleepers on the London and Birmingham. In this case, both men were acting under a misconception: both believed that a totally rigid track was necessary for smooth running, whereas in fact there needed to be a certain amount of give. It is difficult for us now to see just what a state of flux characterized railway building at this time. Brunel himself described the situation:

> At the present moment several eminent engineers, practical men and having experience in the subject, differ very much in their opinion not merely in the best form of rails but apparently even as to requisites. Mathematicians have been called in, in the hopes of settling these differences by theoretical investigation

> apparently without much effect, influential bodies amongst the subscribers of some of the companies have taken part in the discussion and the consequence is that in the various railways which are now in progress different modes of construction are being adopted, the parabolic or fish-bellied rail, the parallel rail – rails with the chairs or supports at every three feet – others with the chairs at 4, 5 and 6 feet apart – and others on the contrary with a continuous support of masonry . . . and besides all these actual experiments making upon such very large scales – there are schemes and inventions without number almost daily brought before the public.[14]

History may have proved Brunel wrong, but his decisions were not necessarily prompted by a desire to be different from everyone else. Given his notion of self-contained railway regions, they made sense: it was the basic premise that was wrong. The same applies to the disastrous atmospheric railway. The idea was simple. A long tube would be built between a series of pumping stations. As air was pumped out at one end, so a piston would be pushed along the tube by the air pressure at the opposite end. A flange at the top of the piston poked through a slit in the tube which elsewhere was kept closed by a flexible flap. Attach trucks and coaches to the flange and you had a transport system. It was not Brunel's original idea – it had been tried successfully at Croydon, an atmospheric railway ran through the suburbs of Paris for a decade, and an Irish line was considered quite a success. It was even proposed to build an atmospheric mountain railway in Austria. It was a bold venture, but one where the technology was simply not available to carry the idea successfully into practice. The leather cover over the slit rotted in the salty air, and the system ground to a halt. In theory it offered the possibility of silent, pollution-free travel: in practice it proved to be Brunel's biggest mistake. It also reflected the fact that, in railway terms, he was a brilliant civil engineer but made no contribution at all to mechanical engineering.

That Brunel was a difficult, sometimes impossible, man to work with no one has ever really doubted. F.R. Conder worked under both Robert Stephenson and Brunel and summed up the two men.

Stephenson believed in using the chain of command: 'the order of his office was not such as to overburden the Engineer-in-Chief with details that fell properly within the competence of the Residents, or even Subs.' Not that the great man was not capable of suddenly pouncing on some detail to show how little escaped his eye. Brunel was very different. His assistants 'appeared to regard themselves less as the officers of the Company than as the channel of the will of Mr Brunel.'[15] On the other hand, it is very clear that the two men regarded each other as equals, whether disagreeing over engineering practice or rushing to each other for support. These two quotations are from letters from Stephenson to Brunel: the first setting out plain disagreement over the plans for the broad gauge, the second written when Stephenson was working at Menai.

> I find it quite out of my power to [submit] a report on your permanent road – I have written to London and declined to do so. As my opinions of the system remain unchanged, you will I am sure readily see how unpleasant my position would be if I expressed myself in an unequivocal manner in my report; and to do otherwise would be making myself ridiculous since my opinions are pretty generally known.
>
> Don't be alarmed that I may think you officious – my feelings are of a very different character – I am deeply obliged to you, for the interest and kindness you have evinced and I rely upon you being with me a day or two before we move.[16]

There is more than enough evidence to show that Brunel felt that everything in which he was concerned should bear the stamp of his personality. The proprietors of the railway company were there to be manipulated and he tested them as carefully as he tested a new rail or a design for a bridge:

> I think I gain ground with Mr Mills – he seems an amicable man but pig headed; Fenwick I think is a friend – Gibbs will go with the Bristol committee – Bettington a jobber but probably

> caring little about anything but his salary & shares - Grenfell must be humoured - Garver very doubtful stupid enough & proportionally suspicious - Hopkins I hardly know - Simmonds a hot warm temperature just such another as R. Claxton - ie warm friend but changeable always capable of being a devil of an opponent - Wylds I don't know - and B Shaw - Mill's tool - In the whole Mills is the leader of the working men and will generally manage to have the committee - upon him therefore I will regularly hang myself.[17]

Brunel was the supreme example of the engineer as visionary. Where others might cheerfully accept the job of chief engineer for the line from A to B, he looked beyond its immediate boundaries to see it as part of a greater enterprise. This could lead him into errors - errors which cost others, notably shareholders, dear - but it could lead to triumphs. It was just such a vision of a railway that convinced him that the Great Western should not end at Bristol but should continue on across the ocean to America. This led him into ship-building and to designing great ships. No man is perfect - and no one ever claimed perfection for the impatient, fiery-tempered little engineer in the big hat - but Brunel's imperfections were never those of the small-minded. He was a man of excesses. He took too close a personal interest in every project in which he was concerned, driving himself beyond all reasonable human limits. He paid too little attention to the mundane practicalities of the real world of compromise and the balance sheet, but he gave to railway building an heroic dimension for which history has loved him. A world ruled by Brunels would have been chaotic; a world without him would have been an altogether duller place. From his first visionary dreams of a great wide trackway carrying smoothly gliding trains to his last great achievement, the stupendous bridge across the Tamar at Saltash, he retained his originality and individuality. Anyone reading his diaries and correspondence must be struck by the seemingly effortless alternation between concern for minute detail and a sudden opening out of the mind to take in distant vistas of what the world might become. Much of the history of railway building is a tale of slow, painstaking advances but in the early years men of genius took sudden leaps into the unknown. George and Robert Stephenson grabbed the concept of the steam locomotive as the force that could make railways live, and forced others to accept the rightness of their judgement. Brunel was to bring real advances to the civil engineering of the lines – his extraordinary bridge of flat, brick arches at Maidenhead, the great tunnel at Box, the revolutionary bridge at Saltash. But more than anything, the Stephensons and to an even greater extent Brunel fired public imagination. They transformed railways from purely utilitarian transport systems to the world of romance. It is no accident that even today, a century and a half after work began, the mere mention of the initials GWR will bring the enthusiasts scampering. Many transport systems have been admired: the GWR was, and is, loved. And that love comes in good part from an identification with the men who built it – the chief engineers.

This chapter has concentrated on the best-known figures in the whole of railway history, but alongside them were phalanxes of other engineers. Some were to achieve almost equal fame, men like Cubitt, Fowler and Locke; others were to live on only in scarcely distinguishable signatures on gently mouldering documents, or as footnotes to the histories of half-forgotten lines. But all, great and obscure, played their part in the building of the railway system. Each of them imprinted something of his own personality on the lines he built. The cautious might creep round a hill, the daring thrust through the middle of it; but whichever course was chosen the marks were left on the land for future generations to admire or shake their heads over. In the end it was all quite basic, as Brunel pointed out when discussing how his rails were to be laid:

> It appears to be altogether forgotten that altho' lofty embankments and deep Cuttings, Bridges, Viaducts and Tunnels are all necessary for forming the level surface upon which the Rails are to be laid yet they are but the means for obtaining that end and the ultimate object for which these great works are constructed and for which the enormous expenses consequent upon them

are incurred consists merely of 4 equal parallel lines not above two Inches wide of a hard and smooth surface and upon the degree of hardness, smoothness and parallelism . . . of these 4 lines depend the speed and cost of transport and in fact the whole result aimed at.

CHAPTER SIX

The Engineers

It appears from a list relating to the number of schemes for new lines in which the principal engineers are respectively engaged that Mr Brunel is connected with 14, Mr Robert Stephenson with 34 Sir John MacNeill with 37 Mr Locke with 31 Mr Vignolles [*sic*] with 22 Sir John Rennie with 20 Mr Rastrick with 17 Mr Miller with 10 Mr Gravatt with 10 Mr S Hughes with 9 Mr W Cubitt with 11 Mr Gibbs with 12 Messrs Birch with 7 Mr Blunt with 8 and Mr Braithwaite with 9.

Felix Farley's Bristol Journal, 27 December 1845

The list of 'Engineering Engagements for the Session' gives some idea of the demands made on the leading railway engineers of the age and, in its mixture of famous and less familiar names, also gives a sense of just how many men were now involved in directing major concerns. Where did they come from? The engineers who had made their reputations in the canal age, such as Thomas Telford, were already old men when the Railway Age began, so to a large extent the railways gave birth to their own generation of engineers. Some thrust themselves forward as George Stephenson had done, pulling others with them. Joseph Locke was just such a man.[1]

Locke's father was a coal-viewer in various collieries, mostly in the Yorkshire area, and the boy grew up and received his schooling in Barnsley. At the age of 13 he was sent out to work as an apprentice to a land surveyor, but when he found that instead of learning surveying he was expected to work as a domestic servant, he left, with no cash and no food, to walk 30 miles back home across the moors. He then worked in the colliery office for a while and might well have stayed there, gradually moving his way up to some such post as colliery manager, had not an old colliery friend of his father's called in on a visit from the north-east – the friend was George Stephenson. It seems Stephenson must have formed a favourable opinion of the boy, still only 17, for he took him back with him as an unpaid apprentice. It was a decisive moment in his life. The year was 1823, work was beginning on the Stockton and Darlington and Robert Stephenson was about to set off on his South American adventure. To some extent, the protégé now took the place of the absent son, and by 1825 he was competent enough to be placed in charge of a simple line from Black Fell Colliery to the Tyne. He kept up an early correspondence with Robert Stephenson, describing his work on surveying the line from Leeds to Hull and adding:

> Whilst surveying what do you think I did? – only what others have done – fell in love! and (you may be sure) with one of the most enchanting creatures under heaven.

Alas, a surveyor's life was one of constant movement, and the line and true love were left behind. He was soon deeply involved in the Stephenson railway projects. He assisted Robert Stephenson on the Canterbury and Whitstable Railway when Robert returned to England and, more importantly, he worked under George Stephenson on the Liverpool and Manchester. This was not a particularly happy time as he became embroiled in arguments between George Stephenson and the railway company. The company who paid Locke's salary regarded him, not unreasonably, as their employee and expected him to work on their line. Stephenson regarded him as a personal assistant who could be sent off to Canterbury or anywhere else he chose. If the company did not like the arrangements, declared Stephenson, then Locke could go. And off he duly went, leaving Stephenson with personal responsibility for the troublesome Edgehill tunnel. The works soon were, frankly, in a mess with shafts sunk as much as 20 ft. away from the line of the tunnel, drift tunnels advancing towards each other on such inaccurate headings that they would never have met and falls caused by poor workmanship. The company ordered Locke to report on the works. He was in an invidious position, being asked to criticize his patron and employer, but he faced up to the task and gave a fair, if damning, review of the works. It was to be his last action on the Liverpool and Manchester, but not his last embarrassing confrontation with the older engineer.

Locke remained tied to Stephenson, and was given the job of surveying the line of the Grand Junction, linking Birmingham to Warrington, Britain's first trunk railway. He was now at the end of his contract with Stephenson. The directors, faced with a young man on the way up the engineering ladder and an older man whose methods were beginning to show their age, arrived at a truly British compromise: they appointed them both as chief engineers, Locke to have the northern end, Stephenson the southern. Like so many other compromises, it was doomed to failure. Stephenson continued to rely on young, often unqualified assistants – a system which often benefited the assistants more than it did the company. Locke took personal control, and had soon arranged contracts for the whole works on the northern end on very satisfactory terms. Stephenson withdrew, and Locke's career was secured.

Over the years his contribution to railway engineering was to be immense. During his work on the Grand Junction he was faced by questions that troubled so many engineers in those early days: what sort of rails should be used, how should they be fastened down and to which sort of sleepers? In 1834, Robert Stephenson over on the London and Birmingham had invited engineers to supply ideas for rails and supports. They responded with a bewildering array of often bizarre notions. Model 28, for example, called for a system in which 'continuous cast iron rails rest on ballast without the intermediate support of Chair Blocks and Sleepers' while in Model 15 'The rail is supported on slate sleepers by brick arches.'[2] Locke opted for a system, largely of his own devising, in which a double-headed rail, looking in cross-section like a dumb-bell, was keyed into chairs by wooden wedges, and the whole was mounted on wooden sleepers, set at right angles to the rails. With slight modifications, this was to be the standard used in Britain for the next hundred years. Locke went on as a civil engineer to build railways throughout Britain and Europe, and became closely associated with the splendid locomotives that came from the workshops at Crewe.

Almost without exception, the young engineers took the road of practical experience rather than formal education, whether they came from a comparatively poor, artisan background like Locke or from a more prosperous professional home like John Fowler. Fowler's father was a land surveyor in the Sheffield area. He was educated at a local boarding school but, instead of going on to university, left school at 16 at his own request to be a pupil of J.T. Leather, engineer at the Sheffield waterworks.

> Before I was nineteen I was a good engineering surveyor and leveller, could set out works, and measure them up for certificates to be paid to contractors.[3]

He first became involved in railways when George Stephenson began surveying the route for the North Midland. In his search for easy gradients, he took the line from Derby to Leeds in a great arc that missed out such important industrial centres as Sheffield and Barnsley altogether: Sheffield was supposed to be content with a mere branch line. Ironically, Leather had actually appeared as a witness against the Liverpool and Manchester Railway Bill, declaring that on the Weardale Railway, locomotives had been abandoned in favour of horses, and that the future of the steam railway was at best dubious. Now, in the mid-1830s, attitudes had swung so far round that there was a desperate hunt on for a line that would bring Sheffield on to the main line. Fowler joined Vignoles in surveying possible routes in the area, but it was clear that there would be little such work as long as he remained in the Sheffield area. He therefore attached himself to one of the leading engineers of the day, John Urpeth Rastrick, who had built the Stratford and Moreton Tramway in the 1820s and had designed and built locomotives. He was a man much in demand when Fowler joined him as an assistant in 1838 and was able to pick and choose among prime schemes – a man who had no need to kowtow to anyone. When the directors of the Manchester–Birmingham line announced a deviation without consulting Rastrick, he promptly resigned. The first his assistants heard of it was when he came into the office and said:

Well, it's no use drawing any more bridges on the Manchester and Birmingham; we must now attend to the London and Brighton.

Young Fowler hoped to get a rise, but Rastrick said that he had got so many applicants wanting to work on the London and Brighton – 'the crack of all the other lines' – that he had young men offering to work for nothing. Soon, however, Fowler was to win his spurs in an eight-mile line from Stockton to Hartlepool where a new dock complex was to be built. He surveyed the line, faced up to a gruelling cross-examination in Parliament, for there was strong opposition, and in 1841 had the satisfaction of presiding over eight miles of completed railway. The life of the chief engineer was not always glamorous – he did everything 'from buying the engines to shutting the carriage doors' – but the completion of a line, no matter how modest, placed him in a small élite.

Fowler did not move immediately to great works. But by 1846, he had enough confidence to take on the entire responsibility for building the East Lincolnshire Railway. He guaranteed to oversee everything 'from the obtaining the Royal Assent to the complete opening of the whole line'. He was to pay the resident engineers and design the stations, order the rails and pay for the surveys.[4] The company gave him sweeping powers to ensure he could meet his obligations. The engineer 'shall have full power to direct the Contractors to facilitate or push the works . . . and the Contractors shall immediately on such directions being given increase the number of their workmen'.[5] In 1853, he was able to take on the task of constructing the important Metropolitan Railway, which was to be the first link in the complex chain of London's rail system, the Underground. The line was to run from Paddington past King's Cross to Farringdon – no great distance at just under four miles. The engineer, however, was faced with the problem of pushing a route through one of the most crowded areas of London, served by major thoroughfares – Praed Street, Marylebone Road, Euston Road and Farringdon Road. Fowler opted for 'cut and fill', not a new technique – it had been employed in the canal age – but one never used on this scale before. A deep cutting was dug in the streets, then covered by arches of iron and brick, on top of which the street was eventually relaid, and the line was hidden away, the only evidence of its existence being the ventilation shaft. But during construction, traffic had to be redirected down side streets, and pedestrians had to cross the 'yawning chasm' of the cutting on slippery planks. Fowler faced new problems when the line was extended westward. Here much of the line was in open cuttings, but it passed right through the middle of fashionable Leinster Gardens, with its strictly symmetrical terrace of houses. A yawning gap in the middle was unthinkable, so an ingenious answer was found. Two dummy houses – front walls with nothing behind them – were put up over the gap, and there they remain, 23 and 25 Leinster Gardens, as substantial as a Hollywood set. The successful engineer was required to consider more than just the best line for his railway.

Fowler's rise was steady rather than spectacular, but it reached a grand climax in 1890 with the opening of the Forth bridge. It shares with Brunel's Saltash bridge the distinction of being instantly recognizable and of being of revolutionary design. Its design grew out of disaster. It is generally easier to find accounts of successes than failures, but one failure at least is an exception – the collapse of the viaduct across the River Tay in 1879, just one year after its opening. The one reason that so many people know of the event is that it was commemorated in some of William McGonagall's most excruciating verses, which begin:

Beautiful Railway Bridge of the Silv'ry Tay!
Alas, I am very sorry to say
That ninety lives have been taken away
On the last Sabbath day of 1879,
Which will be remember'd for a very long time.

The lines are generally quoted amid howls of mirth, which the verse richly deserves, but there was nothing comical about the event itself, and serious engineers set about applying the lessons learned. Studies were made of wind pressure on long bridges, and one result was that plans for a suspension bridge over the Firth of Forth were abandoned. Instead, the present design, generally known as a cantilever bridge, but more accurately thought of as three diamond-shaped trusses linked by girders, was built. It still ranks as one of the triumphs of nineteenth-century engineering.

The story of the bridges over the Tay and the Firth of Forth highlights both the glories and the tragedies that could colour the career of a chief engineer. By the time Fowler retired, he had a knighthood to commemorate his achievements: Thomas Bouch, the hapless designer of the Tay bridge, ended his career in disgrace. Within a year of the Tay disaster, he was dead, broken in body and spirit. His crime was to cut his costs to the limit, which might have been acceptable, but he compounded his first error by paying little attention to the contractors entrusted with the work. It was a common enough story and one which he himself had warned others against many years before, when he was in charge of work on the Eden Valley Railway. Bouch wrote complaining of a lack of clear directions on construction standards in the contracts. His chief concern was over a viaduct, of which he wrote: 'unless the very greatest care is taken with it I should be very doubtful of its standing'.[6] If only he had heeded his own advice. But chief engineers were busy men, torn between the competing demands of many projects. It was a matter of fine decision and judgement as to how far they could trust their assistants, and the men who were to translate plans on paper into brick, stone, iron and steel. Not all disasters were as widely publicized as that of the Tay bridge, but the causes were often similar – it was seldom a case of genuinely poor design, but rather of corner-cutting, penny-pinching and poor supervision. When the viaduct carrying the Leeds and Thirsk Railway over the River Nidd at Knaresborough collapsed, it was certainly a spectacular fall. The river was blocked and it was said that locals were picking fish carried up by the flood waters out of the streets. A committee was set up to find out what went wrong.

Everyone agreed things had indeed gone badly wrong, but everyone denied responsibility. The contractor, George Wilson, declared that he had always said it would happen – the engineer had specified the wrong sort of stone. The resident engineer admitted he had not kept an eye on things, but pleaded overwork – 'I have had all the masonry to look after, it was more than I could do.' William Cubitt, called in to investigate, agreed that the plans and specifications were fine, 'but at the same time I feel bound to say that I deem it impossible to have carried this design fully into effect according to the detailed plans and specifications, honestly and truly both in letter and in spirit at the sum of money for which the work was undertaken'.[7]

The case was simple: margins were tight, the contractor inexperienced and supervision inadequate. It was no good the chief engineer bemoaning his fate, declaring that his plans were fine if only they had been properly carried out, for that was an essential part of his job, to ensure that they *were* properly carried out. The engineer walked a difficult path: stray too far in one direction and he would fall into the bog of administrative detail, sucked down into a morass of minutiae; wander too far the other way, and though the way was easy he would soon find himself out of sight of the plodders on the main road. No man, not even Brunel, could be master of every detail, though Brunel certainly did his best to involve himself in every aspect of the work. He even went to the lengths, when looking for a route across Wales to link up with what would be a new ferry service to Ireland, of hiring a steam boat and trying the various crossings himself. He reported back that the line from St David's was no good, but he had high hopes of a route from Abermawr.[8] But equally no one could afford to pass too much responsibility to underlings.

Once a railway bill received the approval of Parliament, work could begin in earnest, and by then a command pattern would have been established. At the head was the chief engineer; beneath him were one or more resident engineers, depending on the complexity of the works; and they in turn controlled the sub-or assistant engineers. It was a large enough group for work to be delegated, but small enough for the man at the top to be aware of the virtues and shortcomings of all the junior engineers. Brunel, for one, was very clear on what he expected from them:

> The sub-assistants must be considered as working entirely for promotion – Their salaries – their particular employment – and the continuance of their employment, will depend entirely upon the degree of ability and industry I find they possess – Their salaries commence at £150 per annum and may be increased progressively up to £250, and perhaps in some cases up to £300 – They must reside upon such part of the line as required;

> consider their whole time, to any extent required, at the Service of the Company; and, lastly, will be subject to immediate dismissal should they appear to be inefficient from any cause whatever, and more particularly at the commencement must consider themselves as on trial only.[9]

The terms sound harsh, but if the penalties for failure were extreme the rewards for success were accordingly great – the possibility of advancement up the precarious engineering ladder towards fame and fortune.

The chief engineer was, above all, the master planner. He set things in motion, laid down objectives and timetables for the work and, if he was sensible, explained these in reasonable detail to the company directors who were his employers. Charles Vignoles' report to the Sheffield and Manchester Railway Company is a model of clarity. He begins by pointing out that one essential for success is that the chief engineer must inspire confidence in the subscribers. In the early days, when he was writing, they often worried about the difficulties of construction, sometimes with justification, but often out of ignorance. On this particular line, there were reasons enough to worry: there was the Woodhead tunnel, boring for three miles through the heart of the wild Pennine hills; there were great viaducts striding across valleys; and even where no vast engineering works were needed, the rough ground was going to necessitate severe gradients that would test the locomotives of the day to the utmost. The engineer who could persuade shareholders that such problems could be overcome with ease – or even overcome at all – would be doing a good job. But, as Vignoles agreed, in these matters the interests of shareholders and engineer were as one: 'he must build up his reputation as much, if not more, upon the economy with which his designs may be executed, than upon their merits. In this country, an Engineer's Career in his profession depends mainly on the success of his works as commercial speculators.'

Vignoles went on to lay down his plans. First the line had to be laid out 'with geometrical accuracy', and given the nature of the country he assigned a year for the survey. Then the lengthy business of purchasing the land for the whole line had to begin, negotiating with scores of individual landowners and, if necessary, taking the case to arbitration. In the meantime, work should begin on the great tunnel. Eleven shafts were to be sunk at equal intervals along the three-mile line of the tunnel. The deepest of these shafts was sunk 600 ft. below the surface. He had checked with the metal mines of Derbyshire and was sure these could be sunk at between 40 and 50 shillings a yard. He suggested allowing £500 per shaft, not counting the cost of pumping engines.

> I conceive there will be no difficulty in finding in the Country as many Gangs of Working Miners as will enable the whole number of shafts to be worked at once: first building huts on the Hills for the men; a measure absolutely necessary for the absence of all accommodation for them otherwise.

He estimated sinking at a rate of 7 yards a week, and once the bottom was reached one could start working outwards, creating narrow tunnels or drifts. These would provide an accurate picture of the geology of the land, so that the engineers would know exactly what to expect when work began in earnest. Meanwhile quarries would be opened above to supply stone for the works. All this could be done with the cash in hand, without making any further calls on shareholders. And when the shareholders saw how well and ably the works were progressing, there would be no problems at all when it came to asking for the next cash instalment.[10]

Reading Vignoles' report one is filled with admiration for its lucidity and wholly convinced by its arguments. Alas, it was to bear little relation to harsh reality. Work began in 1838. The company was reluctant to release funds. The weather was atrocious and Vignoles' nicely calculated progress rates were soon revealed as optimistic fictions. The 'absolutely necessary' huts were not built – indeed for a time nothing was built – and it was only with great reluctance that the company agreed to provide tents for the men who had been sleeping in the open. Shareholders, far from overflowing in confidence, began unloading their holdings as fast as they could and the price of the shares plummeted. Vignoles, a man with a keen sense of personal honour, hung on to his own shares – he had by then bought over a thousand £100 shares – and refused to sell when the price rose. He could have saved his money, but he preferred to demonstrate to the world his unshakeable confidence in the line. It was a noble, but ultimately foolish, gesture. The company made a call on the shares which Vignoles was unable to pay and he sought the help of Lord Wharncliffe who agreed to let him forfeit his shares. He would lose his investment, but would have no further debts. The company repudiated the deal: Lord Wharncliffe resigned and so did Vignoles. The work passed to Joseph Locke. Vignoles went on to recover his good fortune, but his career stands as proof that though a chief engineer might acquire riches, those riches could be dearly won. His own skills might be of the highest order, but he could still fall victim to circumstances over which he had no control.[11]

Vignoles was unfortunate in being faced with a particularly difficult board of directors, working at a time of financial crisis and on an unusually difficult line. His ultimate undoing was his personal financial involvement in the line, but this was a commonplace among engineers. When George Stephenson was advised to sell his shares in a rising market, he testily declined, pointing out that he had invested for the good of the railways, not for speculative profits. At some time or other any engineer might expect to be faced with unreasonable employers, difficult terrain or wobbly finances. It was Vignoles' misfortune to be confronted with all three at once. But even had life on the line been easier, it is doubtful if things would have proceeded as logically and as evenly as his simple outline suggested. Nevertheless, it does give a notion of the tasks facing an engineer at the beginning of construction. As work progressed new demands were made on him.

One crucial area lay in the decisions on how the railway was to be set out on the ground. The original plan may have called for a tunnel, but closer examination might show that a deep cutting would be more effective; a viaduct might seem on reflection a better way of crossing a valley than an embankment. It was here that engineers could show their individuality, for there was ample room for disagreement. Robert Stephenson went to look at the Bishop Auckland and Weardale Railway, and did not like what he saw. He particularly objected to the embankments, some as high as 60 ft., and felt that bridges would have been preferable to these massive banks.

Stephenson might also have added that banking was an expensive business: the cost increased dramatically as the height of the bank was increased - double the height of the bank and you quadrupled the amount of material needed, because of the sloping sides. The same general rule applied in reverse with cuttings, where material had to be excavated instead of piled up.

> The only case in which I can conceive the propriety of this principle to be questionable is when the material is sand or Gravel, easily obtained, and the base upon which the Embankments are to be formed undoubtedly sound - On the Auckland and Weardale Railway however these conditions are not only all wanting but in lieu of them you have slippery material easily affected by wet weather and obviously treacherous foundations.[12]

It was at such times, too, that engineers could display their boldness or timidity, with the thought perhaps never too far from their minds that it was the bold gesture for which they would be remembered by posterity. It was an age when anything, it seemed, was possible - engineers were certainly able to carry out works which, in our supposedly more environmentally conscious age, would have led to loud and prolonged protests. Can one imagine what would be said about a proposal to blast away a whole chunk of the famous White Cliffs of Dover? Yet that is exactly what William Cubitt did in building the line between Folkestone and Dover. Even before the cliffs were reached, he had spanned the Foord gap with a spectacular 100-ft.-high viaduct on nineteen arches, and beyond that there were to be three tunnels, totalling just over two miles in length, and a massive sea wall. But the most daring plan involved the removal of a large part of the coast. Look on a large-scale map drawn before 1843 and you will find Round Cliff Down. It no longer exists: Cubitt decided to blow it away, and called in the army to help. More than eight tons of gunpowder were packed into bore holes and joined up by electric fuses. It was such a spectacular event that tickets were sold and crowds of spectators gathered on the cliff top in January 1843. They were not disappointed:

> A dull, muffled, booming sound was heard, accompanied for a moment by a heavy jolting movement of the earth . . . The wires had been fixed. In an instant the bottom of the cliff appeared to dissolve.[13]

Within minutes an estimated million tons of English scenery was spread over an area of more than fifteen acres, and the railway company had saved themselves £7,000 in excavation costs.

Nothing offered a better chance of indulging in the grand gesture than the railway station. Such a building scarcely existed on the early lines. Passengers on the Stockton and Darlington bought their tickets at local inns, and the surviving Manchester terminus of the Liverpool and Manchester Railway could have been built for a small country town. When Brunel met Robert Stephenson to discuss a possible joint terminus for the London and Birmingham Railway and the Great Western, he took a rather poor view of his contemporary's thoughts on the subject:

> It has always appeared to me that Mr Robert Stephenson and most other persons with whom I have discussed the subject have a very limited idea of the degree of accommodation and inducements which must be offered to the public in their Depots.[14]

He certainly showed that he personally cared about this very public image that a railway company – and its engineer – presented to the world. At the Bristol end of the Great Western he opted for a mock Tudor style, which can still be seen in the original train shed at Temple Meads. With its hammerbeam roof it is a masterpiece of imitation medieval, though now ignominiously reduced to a car park; some, at least, of the old buildings have found a more fitting use. It has to be said that not everyone appreciated Brunel's design, and the *Annals of Bristol* of 1870 was later to describe the station as 'the most disgraceful, dangerous, difficult, and impractical in Europe'. But it was at Paddington that he produced the visually stunning station that stands today as one of the great monuments of what one could call engineering as architecture. Matthew Digby Wyatt was the architect who added the frills, but it was Brunel who conceived of the station as a great glass-roofed cathedral dedicated to the gods of steam.

The lesson that the early railway builders taught was soon learned. The directors of the Huddersfield and Manchester declared that although they were 'desirous of exercising the utmost economy', they had been 'guided by the experience of the older companies who may be said almost without exception to have been compelled to increase the extent of their Station accommodation'.[15] The result was a classical building that became the focal point of a new town square, very much an architect's work, with its grand Corinthian portico. From the outside Huddersfield station could as well be the town hall.

In many stations up and down the country, design was left to architects who produced buildings in every conceivable style, from the mock medieval of Battle, considered appropriate for its name and the site of the Battle of Hastings, to the sophisticated Greek Revival of elegant Monkwearmouth. But it was in the grand city stations that architecture and engineering met and combined to dramatic effect. It was here that something quite new was developed, a style of building that might look back to earlier styles for the trimmings but in its essentials was altogether new. The new stations were triumphant examples of nineteenth-century engineering at its finest, as important as the grandest viaduct or the deepest tunnel.[16]

Travellers visiting Newcastle from the south first cross the high-level bridge over the Tyne. Then the lines swing round in a smooth curve under a vast roof of glass. The whole is supported on iron ribs – three spans that curve up above colonnades, also made of iron. The idea of using an iron frame to support huge expanses of glass was not altogether new, but it had never been used on such a scale before – the Crystal Palace of the Great Exhibition had yet to be built. The beauty and elegance of the station are virtues which the architect John Dobson was to show in many other buildings in the city, but the masterly handling of iron must surely have owed a great deal to the engineering skills of Robert Stephenson. It set a pattern that was to be followed in other grand stations. The curved train shed was to be given greater refinement at York, but it is in the London stations that some of the finest examples can be seen. Paddington has already been mentioned, but if one had to choose just one station to stand for the perfect marriage of architecture and engineering, then it would probably be King's Cross.

William Cubitt, who we last met vandalizing the cliffs at Dover, was joined by his two sons, engineer Joseph and architect Lewis, to build King's Cross. It is a masterpiece of functional simplicity: two spans of glass curve over the platforms, one covering the arrivals side, the other the departures. Outside, this arrangement is echoed in the two arches of the façade. The materials are all materials of the age: London stock brick, iron and glass, with no fuss, no pretension, but a real magnificence. Other companies might go in for frivolities, but the Great Northern seemed content to let the structure speak for itself. King's Cross owes nothing to any earlier age – without the engineering advances of the nineteenth century it simply could not have existed. Neighbouring St Pancras is, if anything, even grander, but its magnificence is often lost and forgotten, overshadowed by the romantic extravagance of the spired and turreted hotel which runs along the front of it. It is here that the contrast between the new world of 'the engineer's building' and the old, using the traditional language of architecture, is seen at its most extreme.

For many, the new temple of iron and glass represented the public face of the railways, a face largely created by its engineers. Small wonder, then, that these men were held in awe by many – the men who built the stations, who shouldered aside mountains to set out their lines, who threw their arches across rivers and gulfs – held in awe by some, but not by all. Dr Dionysus Lardner, a somewhat farcical figure in the history of technology, believed that engineers were little better than mechanics, ignorant of the workings of science and mathematics by which all problems could be solved.

> I have every respect for the opinions of Practical Engineers, but I consider that many are merely judicial men, who do not have extensive powers of inference or generalization, which is a matter of Arithmetic, and I do not consider the Practical Engineer to be the originator of the Data, although some may claim the highest situation as Scientific men.[17]

This was the man who 'proved' by scientific calculation, among other things, that a train going through Box tunnel would 'deposit 3,090 lb of noxious gases incapable of supporting life' and that it was quite impossible for any steamship ever to cross the Atlantic. The engineers could point to real achievements to support their case.

The job of the chief engineer was not, however, limited to the grand and the impressive: a great deal of time was taken up with the ordinary and the mundane. Even the most conscientious chief engineer could not look after everything himself; he depended on a complex system of administrators and assistants to see the work through.

CHAPTER SEVEN

Administration

> I cannot express my feelings stronger than by saying that the success of a measure of this description and the ultimate welfare of the Company depend more upon the character of the Secretary than upon any one point, nothing can be more erroneous than to suppose that a Clerk is all that is required. A Secretary must in fact be able to assist the directors in his judgment and opinions as will allow him to act frequently upon his own responsibility. I should go so far as to say that an inefficient Secretary might be more injurious to a Company even than an inefficient Engineer.
>
> I.K. Brunel, Private Letter Book, 28 March 1836

Brunel stressed the importance to the company of having the services of a competent secretary. Others went even further, and demanded that he should be a paragon of all the virtues.

> The secretary should be a man of firmness and nerve, with conciliatory and gentlemanly manners; of a strong habit of body, able to rough it out in travelling, and possessing a stack of scientific and mechanical knowledge. If he is a draughtsman so much the better, and he should have been habituated to command large bodies of men, and be able to make a public speech at a short notice.[1]

Could such a person exist, and were so many skills really necessary? Surprisingly he could – and yes, they were. Reading through the day-to-day reports and correspondence of a man like Richard Creed of the

London and Birmingham, one is struck by the range of problems that he faced every day and his supreme self-confidence in dealing with them, and indeed with all aspects of railway construction and management. He was eventually to become a director of the London and North Western Railway. One could arrange his correspondence into neat categories, but a much clearer idea of his everyday working life is obtained by looking at it chronologically, and seeing how he continually switched from one topic to another. The following brief extracts and summaries give the flavour of the whole and cover the year 1834.[2]

In January, Creed was much concerned with such mundane matters as organizing supplies of maps and attempting to arbitrate between two factions over which bank should handle the company funds, and was kept busy preparing for the half-yearly general meeting. Soon finance and shares were dominating his thoughts. Initially, scrip had been issued and certificates made out for every registered share, but in the meantime these bits of paper had changed hands many times over. He decided that it was easier to issue new shares than to endorse transfers, but then he had to be sure that calls had been paid up. It was all very troublesome and by March he was being asked to endorse transfers on the share certificates themselves as well as keep a record in the office. This was too much: he drew the line at getting the certificates sent in for endorsement, an act which would have 'irritated our proprietors by asking for what can under no circumstances be of any value to the company'. Mr Creed was no friend of mindless bureaucracy.

March saw an interesting range of activities: he was advertising contracts in the Liverpool and Manchester papers and was reporting good progress in land purchase (most importantly, at prices below the company's estimates) but was still plagued with share paperwork. April finds him discussing the technical merits of the Liverpool and Manchester Railway locomotives, and making the very sound point that no blame could be attached when engineers got estimates wrong since they were pioneers in the field. Such support must have been very welcome to engineers who were more used to receiving blame than understanding, though there is, alas, no record of a response to Creed's notion that one day the steam engine would be replaced by gas. After that he turned his attention to the employment of contractors:

> From what I have seen of Jackson & Shedden and the pains they have taken to make themselves thoroughly masters of the business which they have undertaken, coupled with what I hear in other quarters, I am disposed to think that although they have not figured in any great Canal works as Contractors they are equal to the task. They speak in terms of the highest praise of the specifications and plans which they pronounce to be the fullest and clearest of any which have hitherto been prepared for contracts in their part of the world.

His optimism on this occasion turned out to be misplaced, for they soon showed that they were far from being masters of the works, and had simply not got themselves organized – as Creed put it, 'the wagon was ready but not the wheels'. Even so, he warned against being too quick to invoke penalty clauses and risk losing the contractors altogether. But his understanding and sympathy were ultimately wasted, for by the end of the year they had given up their contracts. He took a similarly pragmatic view over shareholders who failed to pay up on time. Provided they paid up in the long term, nothing was gained by forfeiting the shares, 'but it will be policy not to proclaim that the intentions of the Directors are *so* lenient that their indulgence should be considered as a precedent for the future'.

Creed was forthright in his attitudes, a man who believed in plain speaking. May found him standing firm over one outrageous claim for land purchasing: 'Mr Weall has been too much our enemy from the first not to have made our refusal a pretext for further annoyance.' And during the next month he went on, in quick succession, to look at designs for rails and sleepers and write a scathing letter about a visit from a committee member ('either he has not the talent of making himself understood or I am wanting in the power of comprehending for I certainly do not to this moment understand the sense of the minute') and ended a flurry of letters with a note on his own delight that share dealings were low – an indication that investors were holding on and relying on the completion of the works for their profits.

The secretary showed himself, on many occasions, to be more than competent to discuss engineering matters. After passing on compliments about Stephenson's arrangement of the works, he added a warning note: 'I look with some anxiety nevertheless to the operations in the London clay and am pleased to find that there is on the part of R. Stephenson no such overweening confidence in his individual resources as may lead him to disregard the dearly bought experience of those who have dealt in London Clay before him'. This was a man of experience, in his fifties, discussing the work of one who was, at the age of 30, in charge of one of the greatest engineering projects of the age. Not that age appreciably slowed Creed down, for in August we find him spending a Saturday looking over the workings – a walk of 21 miles.

Throughout the regular correspondence on financial shares, one finds Creed always ready with help – and practical help at that – for the engineering staff. Time and again he shows himself as forward-looking. He points out that without good ballasting a railway cannot be made secure, and the land can supply it. 'We have an *invaluable* storehouse of chalk and flint in the great chalk range on which we are working', and Stephenson is not to worry about where to keep it – the land will be found. Looking even further ahead, he begins to consider the problems that will be created once the line is completed – in particular, the difficulties of running slow goods trains and fast passenger expresses over the same tracks. Finally, at the end of the year, he adds his contribution to the discussion on the competition to supply rails for the works. In a letter he gives firm views on the notion of competition, an interesting note on Stephenson's character and a strong hint of his opinion of the gentlemen of the committee who employ him.

Of the rails he writes:

> As always happens in Cases of this sort by far the greater proportion have been furnished by persons who are more desirous of obtaining the premium than qualified to contend for it.

There were some good ideas amongst the plans for rails, 'although the younger Stephenson with a pertinacity worthy of his father, but the motives for which I respect infinitely more, is unpersuadable as to the possibility of his plans being improved upon'. He suggested calling in experts to lead the committee, otherwise 'the sight of fifty different pieces of mechanism would be only calculated to bewilder'.

All this occupied the attention of one man in one year, and could be repeated for other secretaries, in other places at other times. The secretary, it seems, did have to be a polymath after all, ready to switch his concerns from bank rates in Liverpool to flint for ballasting at a moment's notice. The letters, however, only give a hint as to the nature of the work.

Perhaps the most important job of the administration was that of looking after the finances. Shareholders were notoriously unwilling, especially in the mania years, to keep up with the calls. Naïvely, they believed that having bought shares, all they had to do was keep their beautifully printed certificates in a drawer and bank the dividends. They were simply not prepared for a letter announcing that they were under an obligation to provide more funds. Many ignored the calls, much as they might a begging letter from an unfortunate, but unloved, relative. They then had to be coaxed, cajoled and, if necessary, forced to come up with the money. Not all did, in which case the shares were forfeited, but this did not help the company finances. Many a company struggled on from financial crisis to financial crisis; some simply collapsed under the burden of debt, to be abandoned or swallowed up by a more powerful neighbour.

The opposite side of the cash problem was keeping control of the money going out. The tedious job of book-keeping went to the office staff, who had to record every single item. The same page might record the purchase of eight cane chairs for the office at £2 14s. 0d. and the sum of £2,223 19s. 0d. for buying land; £474 to the lawyer who helped get the bill through Parliament and £1 16s. 0d. for a hat and coat stand.[3] Looking at the diary of a young man who described himself as 'Elected Secretary, or rather, book-keeper', we find a cheerful jauntiness very different from the measured tones of a company secretary. He took a wry interest in everything about the line, but was no great respecter of persons, particularly, it seems, his superiors. His diary begins with work on the line going along very slowly, largely for lack of men and with not much chance of employing any – 'Thirty or forty saunterers about asking 3s/6d per day. Some navigators working 7 to 5.' A few days later he notes 'barrows are upside down – no men at work'. He seems equally unimpressed by his own tasks, which range from 'Making harlequin map of railway' to 'Walk down the line and count the trees'. To which he adds 'We are always doing things twice for want of system.' The same lack of system seems to have extended to the engineering department. 'Mr Lister tells me there has not been two bridges built according to drawings, but all have been altered and some twice over – *Arch* springs from Stevenson's [*sic*] milepost. Sagax!!' The view from the office stool was not perhaps always as complimentary as that set down in official minutes. He also comments on the business that was to occupy so much of the administrators' time – compensation and land purchase. He quotes a claimant's view of the railway – 'Talked about busy meddling folks would allow naught for people's property' – and puts in a grumbling note about having to work late delivering subpoenas for a compensation hearing. And he records the cross-examination of the company representative who had put him to so much trouble with a certain glee – like a 'dentist drawing doubly four fangs'.[4]

Land purchase was a serious matter, but before it could begin, the survey carried out before the bill was presented to Parliament had to be made all over again. This time, however, it was for real: this was a survey which marked out the ground where the work would actually be carried out. On the surveyor's estimates, land would be purchased and work contracts negotiated. If the preliminary survey was important, this survey was doubly so. Now, however, the surveyors were acting with the full authority of Parliament. Landowners might bluster and obstruct, but the railway had the law very much on its side. The rights of the surveyors to go anywhere they chose along the designated line seemed particularly strange in built-up areas:

> Not rarely has it occurred for a respectable inhabitant of Mornington Crescent to find the peace of his family disturbed by the announcement, that strange men were clambering over his garden wall; and on sallying forth, indignant, to demand the reason of the intrusion, to find them coolly engaged, with hammers and cold chisels, in boring a hole through the wall of his tool-house or summer-house.[5]

This particular surveyor found on one occasion that the only place he could set up his theodolite was on the flat roof of a schoolmaster's house, next to the classroom where the pupils were being instructed in the three Rs. Outside the window, a ladder appeared over a wall and down it came the young surveyor who, out of a sense of propriety, descended facing outwards, so as not to stare into the house. The master was outraged, and his rage was not the least improved on discovering that there was nothing whatsoever he could do about it. The children's reactions were not recorded but can be imagined. Whatever the legal rights of the surveyors, there were still attempts to disrupt the work.

> I am very anxious to fence through all Close's land, as he will not allow a cart in any of his fields. I shall therefore feel obliged if you will once more lay out the centre line through the oats and wheat as I could not find a peg today, and I will see that the old scoundrel does not again disturb them, without paying dearly for his misconduct.[6]

On occasions, a company was forced to take drastic action to avoid disruption of the works. The Great Western had to put this advertisement in the local papers.

> Five Pound Reward
>
> Whereas some evil disposed person or persons have cut and taken away certain ropes affixed to the flag pole on Pile-Hill and pulled up certain stakes placed by the company's engineer for the purpose of marking out the line of the Bristol and Exeter Railway; this is to give notice that Five Pounds reward will be paid to any person giving such information as may lead to the detection of the offender or offenders.[7]

The actual work of the surveyor was the same now as it had been when preparing for Parliament, but attitudes were very different. Brunel had counselled complete frankness in order to get access to property. Now he had very different advice to offer.

> The Landowners or their Agents have adopted the course of pretending to see difficulties and objections in everything proposed. If the railway is to be on an embankment it obstructs the view or stops the drainage, if in a cutting it dries the ground up and they would prefer embankment. In fact in the absence of real agreements every point of information which is offered is laid hold of as an objection, and however contrary to your usual practice, you must endeavour to give as little information as possible. We are bound by our act to make culverts and drain the ground as well as before the construction of the railway and therefore do not when you can help it stipulate for particular culverts as they may not at all answer for other drainage.[8]

He goes on to emphasize the great expense of building accommodation bridges, and points out that any hint given to a landowner might be used at a later hearing. He himself set a good example in avoiding disputes and costly claims by giving in to landowners, particularly if to do so did no harm to the line anyway.

> The line was carried as far North as it could be without actually destroying an Ornamental Cottage and Fishing House and Gardens belonging to Miss Payne. I have reason to believe that this lady would not object to the line passing altogether on the other side of the House. This reduces the curve & saves the bridge being built on the skew!![9]

Brunel himself wanted nothing to do with sorting out land holdings, ownership problems and deals, a job much better left to a solicitor. Other engineers agreed, and Thomas Bouch even suggested that a local surveyor should go along with the company surveyor because the land 'is divided into a great number of small patches belonging to different proprietors and which patches have got no visible boundaries and anybody but a practical land surveyor is sure to go wrong in identifying them'.[10]

The surveyors had the right to go on the land, but until the railway company completed the purchase they were the only ones who did have that right and no work could begin. Landowners soon realized that 'when you have *possession* you have *everything* (as the Ladies will tell you, if you question my assertions) – There are many points to be finally arranged between the Trent Valley Company and myself, before you have any right to possession'.[11] Landowners also realized that delays cost the company a great deal – and cost them nothing. They could, they felt, hold out for the most outrageous terms. On the whole, the companies tried to take a balanced view.

> The Directors desire that a fair and liberal compensation should be given in respect of any *real* damage which the formation of the Railway may entail on the Owners lessees or Occupiers of the land required to be purchased, but they are determined to resist as far as they can the exorbitant claims which are now so frequently got up as Railway Compensation Cases.[12]

Arguments over land purchase reverberated throughout the railway construction period, the haggling sometimes reminiscent of a carpet-dealer's stall in an Eastern bazaar. There were good deals struck on both sides, but on the whole landowner and railway company came to terms that were mutually acceptable. Inevitably, there were grumbles from those who found that the railway affected their property, when they had no rights over the actual land on which it was built. A gentleman might have created a park in order to enjoy a view of distant hills, only to find that his view was to be regularly interrupted by a clanking engine billowing out smoke. But as long as the railway kept clear of his land he was powerless and, perhaps sometimes rather more to the point, compensationless.

Inevitably, it was the abuses of the system that attracted the most attention, but it is as well to remember that these were but a few among vast numbers of dealings. The notes for the surveyor on the London and Brighton give an idea of the kind of detail that was required in settling claims.

> Show particularly the Boundary of each cultivation and put down Clover, Rape, Stubble as the case may be. Where there are buildings and Gardens it will afterwards want plotting to a large scale, therefore be particular.[13]

Often the company would be paying for just a part of a single field, and for some it represented a positive windfall.

An assistant on the London and York Railway was able to report more than mere co-operation from one local farmer: 'I was fortunate enough to see Mr. Edwd. Peacock who assents in every sense of the word – in fact he wishes the railroad would pass over every farm he has.'[14] The surveyor must have wondered if he had not offered altogether too high a price to rouse such enthusiasm.

Looking at detailed maps, the whole thing can be seen as an administrative nightmare – so many owners, so many different attitudes and so many different demands. The one thing that was quite certain was that whatever line was taken, there was no chance of pleasing everybody. In 1861, the North Eastern Railway began pushing a line through Wharfedale, only to face an angry pamphlet from a local bigwig declaring that they had chosen altogether the wrong line by not coming to the local limestone quarry. 'We are told', declared the writer in high dudgeon, 'that Companies have no conscience; and the North Eastern Company seems determined, in every part of their conduct towards us, to verify this saying.' But the company gave as good as – indeed, a good deal better than – it got:

> Where were all your great Landowners from December 1859 to November 1860, during the preparation of the present scheme of Railway? Absorbed in the several duties of your position, overwhelmed with the cares of property, or drowned in the various pleasures of field sports and the amusements of an unruffled rural life . . . Where is that man amongst you who then tried to promote the interests of your order by securing the best line of Railway and this cheap Line? Let him now stand forward, the champion of his class, and not attempt to cast blame upon the Chairman and Directors of the North Eastern Railway Company, and those who have exerted themselves to procure for our Valley a good Line of Railway because, forsooth, they have forgotten the great Landowners who could not take care of themselves.[15]

Fixing the route and buying the land gave the administrators no end of headaches, and the problems did not go away once the land was purchased. The long chain of command from administrators and engineers, down through resident engineers and assistants, and contractors and sub-contractors, to the men with picks and shovels meant that inevitably things went wrong. Work was started before negotiations were complete; land was taken that should not have been taken; property was damaged, roads torn up and a host of small problems created. The problems were small when set against the vast scale of railway operations; but these same problems seemed immense to the individuals affected. It might have appeared a matter of little importance if a minor muddy lane was blocked by spoil for a short time, but if that lane was the route that a farmer's cows took between the fields and the milking shed it was a disaster to him. These troubles were all heaped on the hapless head of the secretary, the ultimate problem-solver. Sometimes you can feel the indignation crackling off the page when some local resident, who expected life to continue in the peaceful way enjoyed by his forefathers, writes to protest at an appalling intrusion into his settled way of life. A certain Richard Jackson wanted a bridge to reunite his land, but having got across his main point he seemed quite unable to stop himself running on, getting more and more worked up as he wrote.

> I wish to have it exactly in the place where the old road is which you are intending to alter a good deal to my inconvenience. Your men have taken the soil of a part of my old Grass field for a road and left it in that manner. I beg you will make me a recompense for it. You have done a serious damage in the same field by making a sluice for water & left it in a very unfinished & shameful manner. I think you wish to quarrel about it . . . I cannot think of submitting to be used in such a manner.[16]

Tenant farmers could often call out bigger guns. In the following case, the landlords were the powerful Ecclesiastical Commissioners and this was the stern message they sent.

> You must really look after the Engineer and Contractors hereon or a Summons before the Magistrates will be the inevitable consequence. They have made the embankment without leaving the

> tenant any communication between his house and land of any description, they have pulled down his fences & his stock is left to stray in whatever direction they like. I do hope you will see to this, as Mr Summerson has been very lenient, and *he* was *here* very much annoyed on Monday.[17]

One can be fairly sure that such letters received the secretary's prompt attention.

It sometimes seems that the secretary's office was little more than the company's complaints department. Not only was there the general public to deal with, but the internal quarrels of the company all seemed to fetch up there as well. The engineer Thomas Bouch had quite a tetchy correspondence with the office when he was in charge of the Eden Valley Railway. The trouble began even before work started, when advertisements were placed in the press for contractors to take on the work. This annoyed Bouch, who found himself 'much pestered' by contractors. When they called in they found there were no plans ready and so nothing to bid for. When contracts were let he was no happier, because they were often woefully short of details, if not downright wrong. Bouch described a meeting with one of the contractors.

> He surprised me by stating, that he claims to build the Stations not according to the lithographical plans which I sent and which were exhibited to the other Contractors but according to some most disgraceful tracings which he says he got from your office.[18]

Ultimately the success of the operation depended on close co-operation between the administrators in the office and the engineer in the field. Even the most efficient secretary depended on getting the right information at the start and then receiving regular and accurate reports of how much progress was being made. When none of these conditions was met trouble inevitably followed. The Brandling Junction Railway was one line where things began badly and got worse as time went on. The engineer was George Stephenson and the resident engineer, Nicholas Wood. It was not one of Stephenson's most successful ventures. The line, joining South Shields to

Newcastle and Sunderland, was primarily a freight line, but the gradients as laid out were too severe for heavy goods. Contracts were handed out on the basis of costings that were hopelessly inaccurate.

> It appears that the estimate was such a one as is usually laid before parliament, a superficial one, and was, by Mr Wood's first step, shown to be founded on incorrect data, as far as the gradients were concerned. The practice of making estimates in this superficial manner is one unfortunately too common to be remarkable; but it is not the less reprehensible.[19]

This was, indeed, a common enough problem, but the committee compounded it by appointing a resident engineer who was not, in fact, resident at all. The directors 'made no stipulation with Mr. Wood, as to the duties to be fulfilled by him, or the time he was expected to devote to the concerns of the railway . . . Mr. Wood, being occupied with the interests of other concerns, which had a prior claim on his attention, could not devote his talent and time to the railway about to be formed so fully as an undertaking of this magnitude absolutely required.'

That an administration should hand over the job of overseeing a major project to a part-time resident beggars belief. The results were inevitable, and the administrators, who should have been keeping a watchful eye on expenditure, found themselves forced, as a consequence of their own initial poor decisions, to pay out more and more cash for the project. Unusually, the building of the line was let as a single contract in 1836 to Ridley and Atkinson, but by February 1837 they had been sacked for slow work and the job handed to Greaves and Grahamsley, at a higher rate. One of the sub-engineers disliked the contracts agreed by the office because 'the company put itself into the contractor's hands and had no control of it'. This proved all too true an assessment. Removed from effective control, the contractors were able to do as they pleased. When digging a cutting, they should have taken the spoil away to build up the embankment. Instead, they spread it across farmland, and then dug up more farmland for the material for the bank. The secretary found himself faced with one bill from a farmer for having spoil dumped on his land and another bill for having soil removed. All this, of course, slowed the work down, at which point the unhappy secretary had to agree to pay premiums to the contractors to ensure everything was ready on time. Such were the problems that could follow from one bad decision by the directors.

Some secretaries survived the whole construction period in good order, and the best earned the respect and gratitude of colleagues. But it must have been a relief when the days of trying to squeeze money out of reluctant shareholders, refereeing the quarrels between engineers and contractors, railway company and landowners, and coping with the complex business of land acquisition and compensation came to an end. At least the secretary and his staff could be certain of one thing: when the work was finished and the line was open, their services would still be needed, whereas the engineering staff would have to go and look for new work and new lines. After what had gone on during the construction years, the job of running a railway must have seemed like a holiday.

CHAPTER EIGHT

The Contractors

Mr Peto's is a peculiar case; he has an immense capital, and can afford to send down to all the different parts of his work without knowing fully what is necessary; he takes very large works, so that he can apply over a considerable extent of country a very first-rate system of account-keeping and superintendence . . . That requires a system which very few men are capable of.

Evidence to the Committee to Inquire into the Conditions of the Labourers Employed in the Construction of Railways, 1846

Everyone that wants some thing of me is looking for it this pay which i will not be able to pay unless you Dow somethink for me.

James Richardson, contractor on the Bishop Auckland and Weardale Railway, undated

Here one can see two extremes: on the one hand, Sir Samuel Morton Peto, who at the height of his powers was controlling a work-force of some 10,000 men; on the other, just one of the many contractors who found that the work they had agreed to do was beyond them, men whose finances were so precarious that when things went wrong they began an instant slide into debt from which they might never recover. Yet in building the railway system of Britain, the work depended just as much on men like Richardson as it did on the great contractors like Peto and Brassey.

Virtually all the construction work went to contractors, though the company might be forced to take over the organization itself in an emergency, as Robert Stephenson had done at Kilsby. But such a step was only taken reluctantly. In general the company men were few in number, with an established hierarchy of chief engineer, resident engineer and assistants. Often the chief engineer would be the man of great plans and the resident the man who saw them through to completion. It was rare for the resident engineer to get more than his pay and a nod of thanks but one resident, at least, was rewarded with public honour. To mark the completion of work on the Newcastle and Berwick Railway, a line which included the famous high-level bridge over the Tyne and the splendid new station at Newcastle, a ceremony was held at the station in July 1850. Robert Stephenson rose to speak:

> I have myself, within the last ten or twelve years, done little more than exercise a general superintendence and there are many other persons here to whom the works referred to by the chairman ought to be almost entirely attributed. I have had little or nothing to do with many of them beyond giving my name, and exercising a gentle control in some of the principal works. In this particular district, especially, I have been most fortunate in being associated with Mr. Thomas Harrison. Beyond drawing the outline I have no right to claim any credit for the works above where we now sit. Upon Mr. Harrison the whole responsibility of their execution has fallen, and I believe they have been executed without a single flaw.[1]

The resident engineer's duties would be carefully set out before he started:

> He is to live in the area and personally supervise the works in the area & to make sure that the specifications and directions are followed by the contractor *in every respect*.[2]

He was very much subservient to the chief engineer, and if changes had to be made he was to give the order in writing, with a copy to the chief. On the other hand, he was expected to use his initiative. If local conditions suggested any way in which the original plans could be improved and money saved, then he was expected to come forward with the suggestions. Each month he had to measure the work done and report on the men at work. Above all he was to make sure that there were no unnecessary arguments over money. The resident engineer was working within strict limits, with only a few opportunities for self-expression. Ironically, these particular rules came from the secretary, none other than the arch-apostle of 'self-help', Samuel Smiles. Fortunately for railway historians this demand for regular reporting means that there is a record of the forces deployed in the works and the problems they met during construction. This particular resident engineer worked on the Leeds and Thirsk Railway, where the major problem was the great tunnel at Bramhope. The reports must have made somewhat depressing reading when they appeared on the chief engineer's desk. In the summer of 1846 the tunnel was being worked from more than a dozen shafts, or should have been, for this was the picture at eight of those shafts when the engineer sent in his first report.

No. 2 – Standing – waiting for pumps
3 – Lost the water a few days ago
4 – Standing in consequence of the breakage of a Crank
5 – Been pumping for fully 2 months at the rate of 338 gallons per minute day and night, without being able to overcome the spring
6 – Drowned out – the Engine will be ready in a few days
7 – In consequence of an accident to the Engine on the 29th of May, this shaft is stopt.
10 – Standing for want of an Engine
12 – The slip of rock has been secured[3]

The problems seem self-evidently to have been engine failure and lack of power in the engines that were working, but blame was also fastened on the 'want of system or management' on the part of the contractor. And clearly organization was needed, for the forces he had to control were impressive. In September, when the engine problem seemed to have been solved and water was being removed at the rate of 1,740 gallons a minute (around 2½ million gallons a day), these were the figures for the contractor's work-force:

Quarrymen:	205
Masons and Builders:	215
Bricklayers:	156
Blacksmiths and Carpenters:	84
Labourers:	166
Excavators and Navigators:	674
Miners with labourers:	429
In All	1,929 men

To add to that, the contractor had 124 of his own horses at work and another 264 on hire. This was clearly a man of means. Equally clearly, his success depended to a large extent on the way in which these large numbers of workers were controlled. And the sums of money involved were immense: for the cutting at the north end of the tunnel alone the company paid the contractor nearly £30,000.

It is almost impossible to give an overall picture of what it meant to be a railway contractor for there was so great a variation. At one extreme was a line such as the Birmingham and Gloucester, where work began in 1837 and all the work was parcelled up into small amounts and given out to specialists, so that one contractor might be in charge of excavation on a section, while another was responsible for the fencing. In all around a hundred contracts were let out in separate parcels, and one's sympathy goes to the resident engineer, F.J.C. Wetherall, who had the task of co-ordination. By way of contrast, the line from Ely to March and Peterborough was entirely let to Peto, who actually brought it in ten months ahead of schedule.[4] Just to add to the confusion, virtually all big contractors handed work on to a multitude of sub-contractors.

The process of hiring contractors began as soon as the Act was obtained and the plans prepared. Engineers and administrators had their own notions of how the work should be broken down into lots and the sort of men who should be awarded the contracts. There was a balance between getting the work done cheaply and getting the work done well. Almost all the engineers were in favour of big contractors:

Those parties who from their means and skill are most capable of executing the work well and with spirit, and from their respectability are most likely to abide honourably by their engagements.[5]

The trouble was that in the early years, and especially in the mania years, there were simply not enough big, well-financed concerns to go round. The result was that even a major concern such as the line from London to Southampton was parcelled out among an assortment of small men:

While the work was easy, while prices and pay remained depressed, while nothing extraordinary occurred, the work was done; but when any engineering novelty arose, the poor contractor was powerless. The smallest difficulty stayed him; the slightest danger paralysed him. He could not complete his contracts; he lacked resources to pay the penalty; the works were often stopped; the directors as often in despair.[6]

When big contracts were on offer, the general tendency was to take the lowest tender. A contract estimated by the company engineer to be worth £85,000 received bids ranging from £68,845 to £93,370, and the lowest was duly taken.[7] Given that company engineers were notorious for underestimating costs, taking a contract at 14 per cent below that estimate must have looked like a bargain. No doubt it was, provided the contractor did not go bankrupt trying to keep to the price. When a contract came up, as long as it was big enough, offers might come in from all over the country. A contract estimated with unlikely accuracy at £399,979 in South Wales brought in three local bids and six from places as far afield as St Albans and Glasgow. In this case the successful bidder was by no means the lowest, but he was a local man from Neath.[8] Sometimes, committees conducted lengthy deliberations over the merits of the offer. Thomas Bouch noted ruefully that while they were talking the price had gone up.

Mr Anderson has informed me that his tender was to be £50,000, so that I was not altogether unprepared for an increase

> over my estimate . . . Mr Anderson states his reason for increasing the tender so enormously above his previous offer is the extraordinary change that has come over the labour market . . . I much regret that we did not accept Anderson's offer at the time he made it. Now that his men have struck for an advance of wages I hardly expect he will be induced to repeat it.[9]

The objectives, however, always remained the same, at least as far as the company was concerned. The ideal was a contractor who faithfully and accurately carried out the engineer's designs, completed the work in the least possible time and was cheap. Such a combination was rarely found. Nevertheless, things usually began with a sense of optimism when the contract was drawn up. This was often a most impressive document, handwritten in a fine ornamental script on parchment sheets big enough to use as tablecloths. The bigger the contracts, the more care went into the specifications. A typical large contract from 1846 covered 1,017 chains – nearly 13 miles of line – for a flat sum of £90,000, but with severe penalty clauses for late work. The contractor had to provide everything himself, except the permanent way equipment of sleepers, chairs and rails. The specifications were very precise: for a major river bridge, ashlar – or dressed stone – was specified, with a minimum width for the courses and a note that there was to be one header for every two stretchers. In the cuttings, they were to finish the job by sowing rye grass and clover. But the company went further than this by laying down rules to protect the workers against some of the abuses to which they were all too often subjected. They were to be paid fortnightly at a place specified by the company and there was to be no Sunday working without special permission. The contractors were also required to pay for any police that might be needed on the line. And on top of that there were curious rules designed to protect a very different social stratum: no steam engines within 500 yards of a mansion.[10] Attempts to be precise did not always produce the expected result. In another contract, the engineer had decreed that no broken stone used for ballasting should be bigger than that which a man could hold in his mouth. On visiting the works with the contractor he found their steps dogged by a shambling, ugly man with an enormous head. The engineer demanded to know who he was.

> 'Oh, him?' said the layer of public way, with a grave designation of the obnoxious attendant with his thumb: 'You mean him? You see, Captain Transom, that is my B-B-B-Ballast Gauge.'[11]

Conditions were no less stringent for one of the most famous of all the contractors, Samuel Morton Peto, for building the whole of the Stourbridge Railway.[12] Here, however, there was one significant difference. The cost was £71,000 but Peto was prepared to take just £9,400 in cash, £35,000 in shares and the rest in bonds and mortgages. For a railway company short of cash this was an offer which was all but irresistible. For a wealthy contractor it was also an excellent deal, provided the shares paid dividends and held their value. In some cases contractors were themselves the active promoters of the line, partly because they thought the line would pay, but just as importantly because it would provide them with work. Sadly for some contractors, their wealth could suddenly appear as no more than unsaleable pieces of paper.

One might have thought that a man of Peto's experience would have known how to make a railway, but each step was described in the most precise detail. For the 'good and substantial' boundary fence, the company specified the wood to be used (larch), the distance between the posts, the size of the posts and rails, and even the depth of the ditch alongside. Work on topping banks had to be stopped in wet weather, and there was a wonderful catch-all clause at the end.

> The attention of the Contractor is particularly called to all the Conditions and precautions (whatever may be the cost of complying with them) which are required in this specification for ensuring an uninjured and perfect formation surface and clean and perfect ballasting.

It was a brave man who could take on such a contract, hedged round with conditions, and be confident that it would pay. In the early years there must have been a good deal of inspired, and indeed uninspired, guesswork on the part of both company and contractors. Of the thirty contractors who worked on the London and Birmingham, ten failed. And those were the giants among contractors, some controlling work-forces of thousands. Yet even so, they still had to rely heavily on sub-contractors and gangers. Peto explained, in some detail, just how he kept control over large undertakings.

His first and most important rule was only to let out work to men he knew personally.

> I always make it a stipulation that the men should be paid every Saturday afternoon; to be paid in the current coin of the realm; and that if in any instance tickets are issued or payments made in any other than the current coin of the realm, they immediately forfeit their contract.

Because he was known to insist on good treatment for the men, he felt that he attracted the very best, but it was still a major operation, with perhaps as many as fifty different gangs at work on a major project, so control was essential. He had an agent in overall charge, sub-agents each responsible for 8–9 miles of line, and time-keepers at 2-mile intervals to record the work done. The sub-agents made up the pay weekly – at 4 p.m. each Saturday – and handed it to the gangers. Peto himself made sure that the gangers got a fair profit for their part, while time-keepers stood by on pay day to look after the interests of the navvies and to make sure none were cheated. Such a system was very far from universal, and many navvies might go a month without pay, one result of which was that the pay-out was followed by an almighty binge that lasted for days. Peto's men got regular pay and regular supplies. The agent would go around the local towns and say:

> at such a place or such a place I shall be paying away 500l. or 600l. a week to the men, and it will be your interest to take care the men are well supplied . . . At one place, I saw several butchers' men crying out 'Who wants a fine leg of mutton?'[13]

The other side of the coin was strict discipline: no long drinking sessions for Peto's men. Anyone not turning up promptly on Monday morning without a good excuse was sacked on the spot. Peto was known for his good treatment of the men, at least by the standards of the day, and he was rewarded by hard work and loyalty. There was a mutual trust between men and contractor, and the same was true of the other great figure of the contracting world, Thomas Brassey. It was notable that if there was any dispute between his agents and the company's engineers Brassey took notice of neither side, but went instead to his gangers for a reliable account.

Thomas Brassey represented one extreme of the hierarchy of contractors.[14] He was born in 1809 into a comparatively wealthy family, and began his working career as an assistant surveyor on the Holyhead road. His employer opened an office at Birkenhead and put the 21-year-old Brassey in charge. He married and might have settled for a peaceful life if George Stephenson had not appeared on the scene looking for stone quarries to supply the Sankey viaduct. Brassey might not perhaps have been able to produce the credentials for the local quarry that one gentleman presented when offering stone for the railway.

> I take the liberty to inform you that the stone used for building the old Bridge across the Tweed at Berwick was got from a quarry belonging to our family, and is decidedly the best in this quarter having stood the test of two centuries and upwards. The marks of the mason's chisel are not obliterated even where exposed to the action of the water and atmosphere every tide for that long period of time.[15]

Nevertheless, he made a favourable impression. He shared a keen interest in the railway and Stephenson suggested that he tender for work on future schemes. Brassey's wife proved equally enthusiastic and ambitious, though she may have regretted it in later years as she moved in the wake of her busy husband, occupying eleven houses in thirteen years, from Birkenhead to Rouen. But Brassey duly acquired his first contract for Penkridge Viaduct on the Grand Junction, and it was here that he first worked with Joseph Locke. It was to be the start of a working relationship that was to be of immense value to both

1. The first successful commercial railway, the Middleton Colliery Railway, running past Christ Church, Leeds

THE

Stockton & Darlington Railway Company.

CLASS A PREFERENTIAL 5 PER CENT SHARES.

No 15415

£25 Share

ACT OF 1838

This is to Certify that Edward Gedman Kirkpatrick of Barham Court, in the County of Kent, Merchant; Lana Saul, of Brunswick, Cumberland; and Charles John Bennell, of No. 2, Eaton Place, Middlesex, Esquire, are the Proprietors of the CLASS A PREFERENTIAL SHARE Numbered 15415 in the STOCKTON & DARLINGTON RAILWAY COMPANY, subject to the Rules Regulations and Orders of the said Company.

Given under the Common Seal of the said Company. Dated the Twenty Fourth day of September in the year of our Lord one thousand eight hundred & fifty eight.

Secretary

2. A Stockton and Darlington Railway share certificate, showing the bridge over the River Sherna at Darlington

3. A meeting of the London and Greenwich Railway Board. George Walter, the principal promoter, is third from the left

4. A cartoonist's view of the Railway Mania: everyone from city gent to navvy wanted to check the latest share prices in *The Railway Times*

5. Surveyors working on the construction of the Forth Bridge pose with their instruments

6. A young engineer, possibly Robert Stephenson, poses with his handiwork, the new railway, behind him

7. Even a minor line had its ceremonials. In May 1890 the first sod was cut to mark the beginning of work on the Metropolitan Railway extension to Aylesbury

8. Above the great tunnels giant machines were set to work. At Kilsby, steam engines worked the pumps, while horse gins moved men and materials up and down the shaft

9. By the end of the century the horse gin had been replaced by the steam winding engine: Belsize tunnel on the Midland Railway extension to London

10. The cavernous interior of Kilsby tunnel, with men being lowered down in the cradle

11. Not an excavation in ancient Greece, but the construction of the triumphal arch at Euston station

12. This photograph (*c.*1868) shows the great unadorned train shed of St Pancras nearing completion

13. A contractor and his staff pose before one of the cellars at St Pancras in the 1860s

14. A contractor's train waiting to take the overseer and his navvies to work on the Great Central Railway

15. Navvies at work on the Great Central

16. A navvy wife poses outside her shack on the Great Central

17. Inside a navvy mission hall

18. The Midland Railway carving its way through the heart of London

19. The typical muddle and clutter of a contractor's yard. In the foreground is a stack of centrings used as frames for arches

20. This photograph of work on the Great Central perfectly illustrates the technique of 'cut and fill': spoil from the cuttings has been used to build the intervening bank

21. The London and Birmingham Railway in Camden Town: a scene described by Charles Dickens

22. Floating the tube into place for the Conwy railway bridge, 6 March 1848

23. A horse-drawn tipping wagon

24. Barrow runs at Boxmoor, London and Birmingham Railway, by J.C. Bourne

men. As long as Locke had lines to engineer, Brassey was assured of work; and as long as Brassey was available, Locke could be certain that his designs would be well and economically executed.

Brassey was, like Peto, a man who realized that success was most likely to come if those lower down the ladder felt they were being treated fairly. Estimating costs was at best a chancy business. A contractor or sub-contractor might view an area where a cutting was to be dug and then find when work began that what he thought would be cutting through soft earth was to be hacking and blasting through solid rock. The company was seldom sympathetic. When a contractor found exceptional circumstances on the Settle and Carlisle line, the only response he got was a curt note: 'such a liability is always involved in the acceptance of a heavy contract'.[16] Brassey understood the problems and was prepared to take up the loss himself rather than let the other man go under. These sub-contracts, usually for specialized work such as masonry or tunnelling, could in themselves be very substantial, running into many thousands of pounds. A successful sub-contractor might one day become a contractor in his own right, just as a ganger might hope to rise to fill the sub-contractor's shoes.

To a large extent, like Peto, the secret of Brassey's success lay in choosing good men in the first place and then trusting them to get on with the job. One recruit came from a chance encounter near Cambridge. Brassey had been inspecting an area of ground, known as 'the quaking bog', over which the Great Northern was supposed to run. 'You can stand upon it', he said, 'and shake an acre of it together.' An acquaintance remarked that in that case he had better have a word with a local Fenland engineer, Stephen Ballard. Brassey took his advice, and found Ballard's judgements so sound that when he left the area Ballard came with him as an agent. It was a relationship based on mutual respect and Ballard left a useful thumb-nail sketch of the contractor at work:

> It was impossible to walk along the line without receiving from him very valuable instruction . . . [he would] regard easy works as beneath his notice; he never looked at them; but if there was a difficult point, as he could see by the section, then there was

something to look at, and he would go and always put his thumb on the sore point.

Brassey was equally quick in making up his mind about his subordinates. A young agent arrived with the first estimate that he had prepared for Brassey and was given a thorough going-over, lasting some fifteen to twenty minutes.

> He examined all the details, I mean to say, that in that short time he turned them all over, and stopped at the difficult points; and in the case of one of the bridge estimates, he went through my figures minutely . . . He looked at the details for culverts, to see how I got at their price, and then investigated the calculations as to the price of the brick-work, of the stone-work and the average quantities of earthwork . . . At the end of the investigation, he said 'That will do.' Never after did he look into any estimate of mine in such detail.

Brassey was in his time to be involved in major projects not only in Britain and Europe but as far away as Canada. British contractors, like British engineers, were in demand for the expertise gathered as pioneers in railway building, though their foreign experiences were not always as beneficial as one might have thought. Robert Turner of Bedford worked in Brazil for many years, but when he got back to Britain to work on the line from Henley-in-Arden to Birmingham he found the heat of the English summer of 1893 too much to bear. He was, he declared, 'physically & mentally incapable of dealing with the business any longer' and gave up.[17]

Substantial contractors were not necessarily secure in their work. Thomas Jackson, in giving evidence to the 1846 Committee on the navvies, described himself as a contractor working on three different projects and employing 3,500 men. He had been apprenticed as a builder twenty-five years ago and had been on the railways for thirteen or fourteen years. He was, he said, a 'practical mechanic'. He certainly would seem to have been a substantial man of business, yet a year later he was in trouble. It was standard practice for a company to withhold part of the fees – often 20 per cent – as a surety against unsatisfactory work, delays and so on. Jackson was working on the Chester and Holyhead line and had to ask the directors to pay over £13,000 in advance so that he could get the work done. He promised in April to have it all complete by August. The directors were dubious but agreed, though they added a stiff penalty clause as a safeguard.[18]

Some contractors came up the hard way. John Sharp of Durham, who also gave evidence to the 1846 Committee, began in 1825 as a navvy and eventually found himself at Keynsham on the Great Western. There he got work with the contractor Ranger as a foreman, starting at 24 shillings a week and rising to 50 shillings. He had charge of 600 men which put him more or less on a par with Brunel's young assistant engineers, who started off at £150 a year. But Sharp 'saw clearly that things were going wrong' and left to set up on his own as a sub-contractor. The sub-contractor had none of the capital commitments of the main contractors who were tied to the works. Sharp described one of them, known as 'Ready-money Tom', who hired horses and disappeared owing their owner £150. He was not the only one to get up to such tricks. John Dean was another sub-contractor on the line:

> From his supposed responsible situation, he had no difficulty in borrowing horses from various farmers for railway work, but at the termination of the engagement for the use of the horses neither Dean nor the horses borrowed could be found.[19]

The small sub-contractors rarely, it seems, set out to cheat the men, but all too often they found themselves simply out of funds and unable to pay.

> The men is coming on me for all there wages and as there is a parte cash in hand for work done since it was measured to be payed on Saderday first I hope that you will take it in concideration so that the men may get there wages for they threaten to Distress me at the justice meeting i was told to pay half of the wages and i have and they are not content.[20]

There was no real solution to the contract problem. Rely entirely on the big contractors and there was a danger that they would establish a monopoly and demand ever higher prices; turn to the small men, and there was the ever-present risk that their resources would prove wholly inadequate for the job. And the problems did not end there. The interests of the company and those of the contractors should in theory have coincided, but they rarely did in practice. The company wanted everything done to the highest possible standard, regardless of time and expense. Too many contractors wanted simply to get on, get paid and get out. The problem was there from the beginning. This catalogue of woes describes the wrong-doings of a sub-contractor on the Stockton and Darlington Railway:

> The destruction of the Company's property by this sub-contractor has not been less than 50 or 60£. He refuses to carry on the work any way but that of his own, deposits the Earth where and as he pleases, and runs the risk of lives and property by wilfully running the waggons at improper speed both down the Inclines and across the Turnpikes at Auckland.[21]

And if the work was done well, it was rarely done fast enough to satisfy the company: too few men at work, too little equipment being used – in short, too little money being spent. How far could you push a tardy contractor? One secretary at least was ready with his answer.

> Recent experiences have convinced me of the folly of delaying too long the proper pressure upon contractors to find the requisite number of men, weapons, engines etc on various parts of the works. In the Eden Valley line the arrears to bring up are frightful . . . 37 bridges (some of them public road bridges) are yet to build besides a vast amount of earthwork to do. In No 1, 2 & 3 Contracts South Durham great loss to the Company is daily going on by reason of the delay.
>
> I am concerned that pressure must be put on Appleby in time or we will be in a mess with his viaducts and large cuttings & what I therefore propose to you is that you should at once give every part of his work your careful attention and define what men, implements, wagons and engines you think should be on every part so as to ensure the line being completed early in 1862.[22]

The saga was to continue, with variations, throughout the long years of railway building. Battles between contractors and engineers – sometimes verbal, sometimes all too real and physical – appear regularly in the written records, but nowhere are they recorded in more detail than in the dealings between Brunel and his contractors. And nowhere does the ambiguous relationship come through more strongly. Brunel was well aware that the success of the works depended, to a very large extent, on the success of the contractors. It was a message which he drummed home to his assistants, and it was a vital message, for in those pioneering days really competent contractors were uncommon. A row developed between the resident engineer of the Cheltenham and Gloucester Railway and a contractor. Brunel refused to get involved in the detailed arguments because they were matters of opinion and personality not of concrete fact. But he was very willing to offer general advice, which is worth quoting at length because it is a model of good sense. We shall see later just how willing Brunel was to take his own advice.

> You must be quite aware that unless a contractor is managed with great care and unless he feels confidence in those placed over him by the Compy the works cannot proceed satisfactorily. When a man like Hemming of no great strength of will and from ill health not over strong in mind or body has in his hands such an important part of a whole line and where the progress of the works depend entirely upon him accordingly the interests of the company require that he should be handled with the utmost tenderness. His losses become the Company's in fact through him and if he is harassed by orders difficult to obey or inconvenienced by the want of money or dispirited by harsh treatment or by what he only fancies harshness work suffers and thereby the Company suffers. Now, Hemmings is peculiarly a case in point. We know that he has not a large capital that if he loses much it will be beyond his means and upon the compy will once paid fall the consequences. If he loses time we are the sufferers, he cannot compensate us. If he should be entirely stopped the loss to the Compy in money would be increased in time irretrievable. He is

the horse we have in harness and upon which we must depend and therefore whatever may be his weak points or his vices or whatever our right to treat him as we choose it is our interest not to overdrive him or to starve him, to be contented with his utmost altho' this may fall short of what we originally calculated upon . . .

You like others in your profession and still more your assistants though you may work hard like a comfortable life & comparative irresponsibility and do not know what real anxiety of mind is. When you have tasted one tenth part of what has been my share you will feel how essential it must be for the management of extensive contracts that a contractor should have his mind at ease and above all that he should be able to calculate with confidence upon all payments and upon the fairness of those whose arbitrary decisions govern those payments.

I think you have not kept sufficient check over your assistants to prevent their occasionally using unnecessarily those arbitrary powers which the terms of the Contract give to the company's agents. I admit that it is difficult to prevent this. It is useless to expect to find in the class of young men who generally form the Assistants that nice discrimination which would enable them to adopt just the requisite degree to induce the best attainable performance of the contract (a perfect fulfillment you cannot have) without undue severity, and to maintain their authority without abusing it.[23]

Brunel could be remarkably patient, as he seems to have been with the contractor William Ranger.[24] By July 1836 Brunel was already concerned about the slow progress at Box tunnel and was asking for extra shafts to be sunk and for the contractor to put on extra men day and night. He suggested advancing £1,000 to Ranger to encourage him to speed things along. A month later there was some improvement to record, but Ranger was building up penalty payments for slow work more quickly than he was getting his advances for the work to be done. In November, Ranger was all but begging for extra cash, and Brunel was firmly of the opinion that no more was due but that it would make sense to pay up anyway – 'make it as it is a great accommodation afforded with considerable reluctance & not to be ever thought of again as a precedent'.

Affairs staggered on for another year, but by August they were reaching a crisis. Controversy, too, raised its head. Ranger was claiming so much work done and the company was disputing the amount. It is impossible now to know who was in the right, but it is worth remembering that Ranger's own foreman had seen 'the way things were going' and had got out in good time. In any case, it seems that by October Ranger had completely capitulated and had admitted that the company could have him sent to gaol, but he still managed to impress Brunel by the 'very manly and straightforward manner' in which he admitted former shortcomings and promised wonders in the future. He also managed to put Brunel on the spot by asking the plain question – what would you do in my shoes? Brunel let him go on, and the works staggered forward, if not entirely satisfactorily. By the end of the year, Brunel was writing:

> I cannot conclude this without stating with respect to the spirit in which the exertions of the Contractor are made that nothing appears done on either of the contracts until the necessity is pointed out, & then though it may be done with perfect willingness it appears done as if in obedience to orders rather than from an anxiety to profit by the suggestions and as I cannot be expected to foresee every thing that may be required if I could feel justified in interfering in the proper province of Mr Ranger.

Now this was more than a little disingenuous, for Brunel was quite happy to interfere in anybody else's province if it suited him, but he was far too busy ever to pretend to exert any sort of control over the Ranger contracts. Much of Ranger's work was to be taken over by the McIntoshes, father and son, and Brunel's relationship with them was not to be much happier. He had fewer quarrels with their rate of progress but there were constant arguments over the estimates for work done. The arguments were ferocious and were to end in a lawsuit before the contractors got their due reward, but they were no more than a somewhat extreme example of arguments that were waged wherever railways were being built. To the engineers, contractors never worked fast enough nor well enough to meet engineers' demands, and then they tried to swindle the company by claiming for far more work than had actually been performed. On the contractors' side, the engineers were constantly changing their minds about what was needed and subjecting them to wholly unreasonable delays by not having land purchased in time, not having materials ready, and so on. And, of course, they tried to swindle the contractors by paying for far less work than had actually been performed. All this was commonplace, but in the war between engineers and contractors, Brunel went further than any other engineer was ever to do. The result was the Battle of Mickleton Tunnel.[25]

The tunnel represented the major engineering feature on the Oxford, Worcester and Wolverhampton Railway. It was to be troublesome from the start, not because it presented any great technical problems, but mainly through being passed on to a series of equally incompetent contractors. The first group lasted a mere four months before their funds ran out. The new contractors lasted longer, but still had to borrow cash from the company to pay for equipment. They staggered on through 1847 and 1848 at a snail's pace – there was a period lasting many months when there was no one in charge of the miners. In 1849 the partnership was dissolved and a new one created between William Williams and Robert Mudge Marchant. Brunel must now have hoped for better things: Marchant had not only been one of his assistants, he was also a second cousin. But he warned the young man that though he might make a future by taking on the contract, he could also end up without a penny. Now it was the turn of the company to run short of funds, and work was suspended, with the contractors being paid only for maintenance of the half-completed works.

In December 1850, when work restarted, there was a new sense of urgency. If the whole line could be completed then the company could start running trains and would begin restoring its finances, but as before the tunnel proved the bottleneck. Williams and Marchant were given one last chance, and in June 1851 they were told to put more men to work – a quite legitimate request – or leave. Then the trouble started.

In July, Marchant's agent complained to the magistrates that a sub-contractor had threatened to take over the works, by force if necessary. The magistrates, fearing a riot, swore in special constables, including R.M. Marchant, scarcely a neutral in the conflict. Brunel arrived on the scene and there was an evening of discussion, argument and negotiation that got nowhere. So next day two navvy armies faced each other at the tunnel, all well armed with sticks and cudgels and well oiled with beer. The magistrates had no choice but to read the Riot Act and the Brunel army withdrew with obvious reluctance. It was only a reprieve. Navvies poured in from the surrounding countryside and there was a dawn confrontation at which the magistrates just managed to hold the peace. Marchant with 150 men faced Brunel with 2,000: there was only one possible outcome. Brunel had won, but in doing so he had risked bloodshed and had come within a whisker of attacking the forces of the law.

In the end it was agreed that Marchant's work-force would be kept on, thereby removing one possible source of future trouble, and Peto and Betts took over as the new contractors. The company and Marchant agreed to put their cash claims to arbitration. Williams and Marchant lost out and, as Brunel had warned, Marchant was faced with financial ruin. It was an event that cast little credit on anyone, though it is not difficult to imagine the build-up of frustration as the apparently immovable contractors went on their dilatory way – it was 1851 by now, after all, and the tunnel had been scheduled for completion three years earlier. The firm action of the magistrates had prevented bloodshed, but Henry Waddington of the Home Office surely provided the most accurate summary in a letter to the railway company directors:

> However clear the right to the possession of property may be the law will not allow the party claiming it to assert such right forcibly and in a manner calculated to endanger the peace.

And with that one of the ugliest incidents in British railway history came to an end.

The Mickleton affair highlighted virtually all the problems that were likely to trouble relations between contractors and engineers. It started when contracts were awarded to men who never had anything like the means with which to see them through. It continued when the contractors failed to supervise the works adequately, or to take due notice of the company engineers' advice and cajoling. This was always a difficult area. Where the two sides were far apart, little could be done. Sometimes the fault for delays in construction lay with the young engineers out in the field, who no doubt found writing reports on progress a disagreeable chore – as did many more eminent engineers. But one at least received a stern rebuke from the man in charge:

> Whatever terms of intimacy we may be upon I cannot in my public capacity show more leniency towards you than to others with whom I am bound to act.[26]

Without adequate information, the engineering staff had no means of knowing how the work was going nor if sufficient resources were being applied to see it through. But the assistant engineers could also become too close. The Taff Vale had to warn an assistant that he had been 'very indiscreet' in becoming involved with 'parties who had money transactions with the contractors' and was told very firmly not to 'familiarise with the Contractors'.[27] There was no doubt that the possibilities of corruption were ever present. But even if both parties were honest, there was endless room for disagreement when it came to assessing the value of the work done. In the case of Williams and Marchant, arbitration came down in favour of the company: in the case of the McIntoshes on the GWR the scales were heavily weighted in the opposite direction. Such quarrels were accepted, however reluctantly, as part of the everyday problems that beset big contractors, but they could sink their lesser brethren without trace. The big contractors could make their fortune. Four names that have already been prominent in this story – Peto, Brassey, Betts and Jackson – were to reappear in partnership, not as contractors but as manufacturers of locomotives, the majority of which were to be sent to Canada. But by 1870, after eighteen years of operation, only one name appeared on the letter-head, that of Thomas Brassey. The other great man of the contractors' world, Sir Samuel Morton Peto, saw his business collapse in the financial crisis that swept the country in 1866. However, he was to live on to a grand old age in comfortable retirement; for many of the small contractors and the sub-contractors beneath them, any fall could take them back to where they had begun - back to that great anonymous army, the railway navvies.

CHAPTER NINE

The Navvy

I am a navvy bold, that's tramped the country round, sir.
To get a job of work, where any can be found, sir.
I left my native home, my friends and my relations,
To ramble up and down and work in various stations.

Broadside ballad, quoted in Roy Palmer, *A Touch on The Times*, 1974

The old ballad gives one very popular view of the railway navvy – a free spirit who turned up at the workings and, if not happy with what he found, moved on to the next. Later verses describe him going off drinking after pay day, having a tumble with an 'old wench' behind the pub and taking her back to the camp for the navvy 'wedding ceremony', jumping over brush and shovel hand-in-hand. It is all good colourful stuff and has become a part of navvy folklore. And there is a good deal of truth in it, but it is only a partial truth. The popular image of the navvy is very much that of the tramping navvy. Terry Coleman in his classic book *The Railway Navvies* defines the true navvy.[1] He had to work on big public constructions; he had to live with other navvies in camps by the line and move on to similar works when one job was completed; he had to eat and drink like a navvy – two pounds of beef and a gallon of beer a day; and it helped if he dressed correctly as well, in moleskin trousers, double-canvas shirt, velveteen square-tailed coat, hobnail boots and felt hat with the brim turned up, with a gaudy handkerchief to add a touch of colour.

The popular image also included the ability to do prodigious physical work while living in often appalling conditions; in return he was dubbed the prince of workmen and earned a suitably princely wage. To all that can be added a fearsome reputation for drinking, fighting and rioting. Certainly, there were navvies in plenty who fitted this image, but there were a great many more who did not.

Contemporaries regarded the navvy with horror, and almost drooled over accounts of his wickedness.

> The Character of Navvies is so well known, that I need scarcely State, that Swearing, gambling, drunkenness & Sabbath breaking, prevail to a great extent among them. And it would have given me great pleasure to have been able to state, that the last mentioned wicked practice was not observable upon the works . . . It is also well known that Ale, Porter &c are sold in very many of the Huts on the Line.[2]

But whether condemning the navvy as a brutish rogue or regarding him as a romantic figure, one cannot begin to get a true notion of the navvy and his way of life without seeing him in the perspective of his age.

The term 'navvy' has become synonymous with the railway worker, but in fact railway work covered a wide range of skills. There were specialists, carpenters, bricklayers, masons and miners, as well as the men with pick, spade and shovel. A contractor in the 1830s broke down his work-force into their different categories and noted their daily rates of pay.

Carpenters and smiths	3/2 - 3/6
Strikers	2/-
Navvies	2/- - 2/8
Ganger	3/6
Horse keeper	2/8
Sawyer	3/- - 3/6
Labourer	2/6[3]

This breakdown gives a hint of the complexity of deciding who could be called a 'navvy' for the list contains a separate item for labourers. But for these contractors, the navvy can be taken as the man who did the hard work of digging and loading, cutting and tipping – work that was not only strenuous but also required specific skills. These skills tended to set the navvy apart from the ordinary labourer. However, to put the navvy into perspective he has to be seen in comparison with others in the land who toiled away at hard physical tasks. By far the most common of these were the farm labourers. The young man in the ballad 'left his native home', but what did he leave behind?

Life in the Victorian countryside was no pastoral idyll.[4] The village of 'old-world' thatched cottages was in practice often an insanitary nightmare. A *Times* correspondent investigating conditions in Dorset described a village street where the only stream was one draining down from the pigsties to collect in foul stagnant pools. He found families with just two rooms to their hovels: downstairs there were a few broken-down chairs, and in the one bedroom three beds for a family of nine, including the eldest son and daughter in their twenties. Worst of all were those tied to farmers by their shilling-a-week cottages. The labourers could be evicted at any time, so arguments over wages and conditions were almost unknown in the area. Even the evils that one would have thought were limited to the itinerant navvies were found, labourers being paid in tickets that could be exchanged for farm produce at prices set by the farmer, who thus became their employer, landlord and monopoly shopkeeper. Against the minimum pay of the railway labourer of 2 shillings a day, rural workers were getting a mere 8 shillings a week. These, it has to be said, were extreme conditions for England. Things were worse in Scotland and Wales, and worst of all in Ireland, where the failure of the potato crops in the 1840s brought famine to the land.

If living conditions were poor, working conditions were little better, and even what we now see as a settled way of life, dictated by the rhythm of the seasons, was not quite as settled as it sometimes appears. There were men like Joseph Arch, the farm workers' leader in the 1870s, who, finding he could not keep his family on what he earned as a labourer, set out on the road as a jobbing worker. Among the tasks he took on were drainage ditches, which involved working up to twelve or thirteen hours a day ankle-deep in muddy water. But by far the most important time for itinerant workers was when the harvest had to be brought in. An old countryman described the process:

> I dessay your've heard of the 'lord', as we used to call 'im? Sometimes he was the horseman at the farm, but he might be anybody. His job was to act as a sort of foreman to the team of reapers – there was often as many as ten or a dozen of us – and he looked after the hours and wages and such-like. He set the pace, too. His first man was sometimes called the 'lady'. Well, when harvest was gettin' close, the 'lord' 'ld call his team together and goo an' argue it out with the farmer. They'd run over all the fields that had got to be harvested and wukk it out at so much the acre. If same as there was a field badly laid with the wather, of course the 'lord' would ask a higher price for that. 'Now there's Penny Fields', he'd say – or maybe Gilbert's Field – or whatever it was; 'that's laid somethin' terrible', he'd say. 'What about that, farmer?' And when the price was named he would talk it over with his team to see whether they'd agree. The argument was washed down with plenty of beer, like as not drunk out of little ol' bullocks' horns; and when it was all finished, and the price accepted all round, 'Now I'll bind you', the farmer 'ld say, and give each man a shilling.[5]

It was not so very different from the system used by navvies and their gangers. Gangers were especially common in East Anglia where women and children were employed. Victorian society became outraged, not because young children were working twelve hours a day and more in often atrocious conditions, but because of the opportunity that the gang system offered for immorality. The Gangs Act of 1867 brought some control to the proceedings, stopping, among other things, the employment of children under the age of 8.

It would be wrong to give the impression that the countryside afforded a picture of unrelieved misery, for this was far from the case, but it did present a scene of poverty and hard work. It is as well, when looking at the way in which navvies lived and worked, to remember that at the same time women and children were out in the fields working all daylight hours in winter, breaking open the hard, icy earth to pull up potatoes and turnips with their bare hands, and often returning at night to a place little better than a leaking barn. But the Sheriff of Edinburgh, giving evidence to Parliament, said that the navvies built huts and the workers who came in at harvest-time slept in barns and 'a barn was a better place than any hut I have seen'.[6]

Where did the railway navvies come from? They did not appear with the railway age at all. The name 'navvy' is itself short for 'navigator', the men who built the canals and river navigations of the eighteenth and early nineteenth century. Even after work started on the Stockton and Darlington Railway there were still major canal schemes under way. Telford's great ship canal was being driven through the heart of the Scottish Highlands, while on the Birmingham and Liverpool Junction, the system of deep cuttings alternating with high embankments was being used to create a system as dramatic as anything in the Railway Age. And the men who dug the canals had already, in some eyes at least, acquired that fearsome reputation which was to carry on into railway building.

> In the making of canals, it is the general custom to employ gangs of hands who travel from one work to another and do nothing else.
>
> These banditti, known in some parts of England by the name of 'Navies' or 'Navigators', and in others by that of 'Bankers', are generally the terror of the surrounding country; they are as completely a class by themselves as the Gipsies. Possessed of all the daring recklessness of the Smuggler, without any of his redeeming qualities, their ferocious behaviour can only be equalled by the brutality of their language. It may be truly said, their hand is against every man, and before they have been long located, every man's hand is against them; and woe befall any woman, with the slightest share of modesty, whose ears they can assail.
>
> From being long known to each other, they in general act in concert, and put at defiance any local constabulary force; consequently crimes of the most atrocious character are common, and robbery, without an attempt at concealment, has been

> an everyday occurrence, wherever they have been congregated in large numbers . . .[7]

That this account was wildly prejudiced is unimportant: the railway navvy's character had been given to him even before he started to work.

By the 1820s, the canal map was almost complete and there were large numbers of trained navvies ready to move on. It made little difference to them whether the great deep cutting at Tring was to carry the Grand Junction Canal or its neighbour the London and Birmingham Railway, the work and the techniques were very much the same. These men formed an élite – men hardened to the work, competent in its special skills. Contractors like Peto, who could pick and choose, sought them out for his gangs and they were welcome wherever there was work to be done. But there were never anywhere near enough of them, and more had to be recruited and trained up for the work.

The first and most obvious source of labour was the area through which the railway was to run. The farm labourer looked at the navvy, looked at his own impoverished life and off he went.

> Hodge, deep-chested and broad-backed, discovers by association and comparison that if he can eat as much meat and drink as much beer as the stranger, he can do nearly as much work, so he sacrifices those parish ties so dear to the ignorant and timid peasant, and takes to the shovel and wheelbarrow.[8]

It was not necessarily a permanent arrangement, and only lasted for as long as railway work proved the better option.

> The men when discharged generally leave the place forthwith, and no doubt go to the agricultural districts where most of them came from as I hear them talk of going home, Railway works generally being at a stand.[9]

Recent work has shown that in some areas, notably the south-west, as many as four out of ten navvies were working within ten miles of their birthplace.[10] It was a different story, however, when it came to the wilder countryside of the Scottish Borders or the hills through which a line such as the Settle and Carlisle carved its way, areas which simply did not have the population to meet the needs of the railway builders. Here navvies had to be brought in from further afield and Terry Coleman's 'true navvy' came into his own. It was not just that these were the itinerant workers, the navvies 'on the tramp', forever moving from place to place as whim or the work demanded, but they were strangers moving into an area which had not seen such changes in living memory. And many of them fitted that other great navvy stereotype – they were Irish.

The one thing that soon becomes obvious is that there is no overall pattern that could be applied to the whole country. The numbers at work varied for a start: there would be twice as many men per mile in difficult hilly country such as the Pennines as were needed on the flat lands of East Anglia. There was also the question of the importance of a line. At its height, the Leeds and Thirsk with its great tunnel at Bramhope had well over 7,000 men at work, at 185 per mile. In 1860, the Eden Valley line, meandering around the hills of Cumbria, with one major work, the Smardale viaduct, had a total work-force of 2 company engineers and 1 inspector cum time-keeper looking after the sole contractor employing 4 time-keepers and 667 men spread out at an average of 30 per mile. Establishing patterns is no easy matter; only the work remained much the same.

The same story is repeated when one turns to look at living conditions at the works. The popular image is of shanty towns, rough and rowdy camps, the sort of place, as one contemporary put it, that one might expect to find in California at the time of the Gold Rush. But this was by no means universal. Not surprisingly, the navvies preferred living in a proper house in a town with all the usual facilities – shops and, of course, pubs – readily at hand, to living out in a crude encampment miles from anywhere. One detailed study of men working on the East and West Ridings Junction Railway reveals much information about the line near Knaresborough.[11] This was the great centre of activity, for the River Nidd was to be crossed on a high viaduct and there were extensive works. It turned out that a quarter of the men were locals, a quarter Irish and the rest came from all over England. Most of them lodged locally, but Knaresborough is a small market town now, and was even smaller then. It so happened that all this took place in 1851, a census year, so we have some sort of record of just how much overcrowding this great influx of navvies produced. An agricultural labourer's cottage, which was already a trifle crowded with four children, managed to make space for nineteen lodgers, including four married couples and three children. The Irish and the English tended to go their separate ways, but an Irish widow contrived to find room for seventeen lodgers. The navvies brought more overcrowding to areas which were often already bursting with humanity – areas with few if any proper drains, where houses were built around dark, airless courtyards and where families were crushed together in airless cellars. When work began on the London and Greenwich Railway, there were problems from the start as the surveyors made their way through the slums of Bermondsey, 'laying out the line of road through the most horrible and disgusting part of the metropolis, and during the period when the cholera was raging in that particular neighbourhood'.[12] It is not difficult to imagine the effect of hundreds of navvies pouring into such an area, for the work here was remarkable – a railway carried on 878 arches across the rooftops and stretching for miles. The workers divided into two groups, the English and the Irish, and kept themselves to themselves – when they were not fighting each other. The two areas became known as English Grounds and Irish Grounds, and English Grounds lives on as a street name close by London Bridge station.

Out in the country things were very different.[13] There were said to be 2,214 Scots and Irish working in the Edinburgh area in 1846. The Scots usually found lodgings with local families, but the Irish lived in huts which they built with wood supplied by the contractor and then roofed with turf. On average the huts measured 20 ft. by 12 ft., divided into two rooms with a central fireplace. The beds were arranged in tiers, with two or three to a bed, and a hut would generally be occupied by twenty to thirty men and their wives and families. Robert Rawlinson painted an equally grim picture of life on the London and Birmingham Railway: some men lived out in lodgings; a privileged few had the use of sixty brick cottages built by the company; the rest were in huts.

> If they had been looked after by the men themselves, they were respectable, well built, good-sized huts; but there was no sort of superintending control over them; the men crowded them as much as they chose; one man would take a room and let it out, or sublet it, to others; and they got filthy and dirty, and abounded with vermin . . . fever and small-pox broke out amongst them.

The women suffered as much as the men, though as so often happened, the chief worry seemed to be that they were living with men without having gone through any official church ceremony.

> Demoralization existed to a large extent among the female population; the females were corrupted, many of them, and went away with the men, and lived amongst them in habits that civilised language will scarcely allow a description of.

The tale was repeated over and over again throughout the country. 'In one bed, you may find a man and his wife, and one or two children; in the next one adjoining, probably a couple of young men; again, on top of them, another man and his wife and family.' One blunt statement commented: 'a humane man would hardly put a pig in them'. But for all the condemnation and the moral outrage, nothing effective was ever done to control conditions, any more than anything was done to improve conditions for the almost equally ill-housed farm labourer. Some individual contractors took a more personal interest in the living conditions of the men. Peto described how he had barracks built, with 'a steady married man' in charge, whose wife did the cooking. The men paid a shilling a week to sleep in hammocks – there were separate arrangements for married men. Captain Moorsom, resident engineer on the Chester and Holyhead, described how in their works the company insisted on the contractors supplying decent accommodation, though the result was not quite what they expected. The contractors duly built 'wooden huts, with slated roofs; but the Welsh, who compose a large proportion of the labourers, do not like to take them, as they say, because they are too fine'.

There was nothing fine about the settlement that grew up on the windswept moors above Woodhead tunnel, as described by a Manchester surgeon.

> The huts are a curiosity. They are mostly of stones without mortar; the roof of thatch or of flags, erected by the men for their own temporary use, one workman building a hut in which he lives with his family, and lodges also a number of his fellow workmen. In some instances as many as fourteen or fifteen men . . . Many of the huts were filthy dens, while some were whitewashed and more cleanly.[14]

They were at their worst in winter, when one man had to dig a path from his hut to the works through 4 ft. of snow. When the surgeon went on his visit, he was even less impressed with the navvies than he was with their homes. They were 'excessively drunken and dissolute – that a man would lend his wife to a neighbour for a gallon of beer – that a large proportion of both sexes [more than half, he stated] laboured under some form of syphilitic disease'.

By the end of the century conditions were much improved at some sites. At the great Severn tunnel works that went on from 1873 until 1886, what was in effect a whole new village was erected. There were houses built to take two married couples and twelve lodgers; other smaller ones for two couples and six to eight lodgers; a number of cottages, quite grand semis, for the foremen and their families; and the principal foreman could enjoy the luxury of a detached house all to himself. There was a mission hall, day school and Sunday school, and the company arranged for local tradesmen to make regular calls with supplies. Any Woodhead navvy who strayed into that settlement would have thought he had died and gone to Heaven.

Of all the settlements, camps and villages in which navvies made their temporary homes, the most famous were those of the Settle and Carlisle Railway.[15] The advance party arrived in the winter of 1869–70 with what can best be described as a hut on wheels, which came to be known as 'The Contractors Hotel', and the men pitched tents on Blea Moor. Soon the works were under way and by the summer of 1870 there were over a hundred huts between Batty Moss and Dale Head, built of wood and covered with roofing felt. Sedburgh's medical officer of health came to look at Dale Head and returned with the usual story of gross overcrowding: 'In one hut I noticed five bedsteads jammed so tightly together that the sleepers in reaching the further beds must necessarily clamber over the others.' In time, a series of settlements developed along the line, often given quite fanciful names: Jericho, Jerusalem, Sebastopol, Inkerman and, with a pleasant ironic humour, Belgravia. The Crimean War names might seem a little bizarre, but many a navvy was as familiar with Sebastopol and Inkerman as he was with Leeds or Birmingham – in the winter of 1854–5 navvies had shipped out from Liverpool to build a railway to supply the beleaguered British army. The largest of the settlements was Batty Green, with a population of over 2,000, and the highest was by the Rise Hill tunnel workings at an altitude of 1,300 ft. These 'villages' were permanent enough for their names to be used in official records, mainly in registering deaths at Chapel le Dale. The atrocious sanitation, or rather lack of it, was the problem. Huts were surrounded by stinking cesspools which could only be crossed on crude plank walks. Rats ran everywhere. Not surprisingly, disease, notably cholera, raged unimpeded through the shanty towns, and it was the women and children who suffered the most. Little of this tragedy found its way into official railway documents. One terse note recorded a contribution of £20 towards 'the cost of enlarging the Burial Ground at Chapel le Dale rendered necessary in consequence of the Epidemic small pox among the navvy population'.[16]

The navvies were a shifting population: in May 1871 there were 6,980 at work at various sites along the line, but many got tired of being held up by the wet Pennine weather and by June over 1,000 had left, just when the company had been hoping to increase the numbers at work. But there were always enough to ensure the settlements were busy places: Batty Green, in particular, became almost a regular village with its own bakery and slaughterhouse and traders coming with food. The surroundings, however, were so boggy that a special cart had to be employed which used a barrel in place of conventional wheels to stop it sinking into the morass. Schools and missions were established but seem to have done rather less business than the local pubs. On one occasion when life was getting a bit dull in the Gearstones Arms near Ribblehead, a navvy chucked a stick of dynamite on the fire – that, it seems, livened things up no end.

Few paid very much attention to the lives of the navvies, let alone their families. One group that did were the police. At a settlement on the Leeds and Thirsk line, there were 93 cottages, offering a total of 245 rooms, home to 531 men, 142 women and 221 children. There were attempts to start a school, and many of the parents were willing to pay, but most simply did not stay long enough.[17] A young Wiltshire woman described just what it meant to leave home and tramp up to these workings. Her account provides a rare insight into the lives of the women who followed their navvy husbands.

> Well, he was just middling steady, and us was main comfortable for most a year; and then 'twas wintertime coming, and they was working nothing but muck. Charley was tipping then, like he is here; and 'tis dreadful hard to get the stuff out of the wagons when 'tis streaming wet atop and all stodge under. Then, you see, he was standing in it over his boots all day long; and once – no, twice – when he draw'd out his foot, the sole of his boot was left in the dirt; new ones, too, for he had a new pair of 15s boots every week. So he cudn't stand that long. One Saturday night he took out his back money, and said us wid tramp for Yorkshire; for he'd a work'd there and 'twas all rock, and beautiful for tunnels. I didn' know where Yorkshire was, but I hadn' never ben twenty mile nowhere. 'Twas four year agone, and I wasn' but just seventeen year old, and I didn' like for to go; and 'twas then us began for to quarley so. He took his kit, and I had my pillow strapped to my back; and off us sot, jawing all along. Us walked thirty mile a day, dead on end; it never stopped raining, and I hadn't a dry thread on me night or day, for us slept in such mis'rable holes of places, I was afeard my clothes 'ud be stole if I took 'em off. But when us comed to Leeds, where Charley know'd a man as kept a public, if I lives a thousand years I shan't never forgit the fire and the supper us had that night. But 'tis a filthy, smoky place; and when I seen it by day, I says, Well, if this is Yorkshire, us had better a stopped where us was, dirt and all. And what a lingo they talk! – I cudn' for the

> life of me understand 'em; and I were glad that Charley cudn' git work he liked there. So us had three days' more tramp – just a hundred mile – to a tunnel. They was a rough lot there; and then us seen and done all sorts o'things I wish I'd never heard on.[18]

The moralists were all too ready to dismiss the navvies and their womenfolk – whom they generally assumed to be unmarried unless proved otherwise – as immoral, shiftless and given to all kinds of vices. They wrote of men selling their wives to others for the price of a beer: few, it seemed, noticed the great heroism of the women who followed the works and somehow contrived to keep their families together. Commentators on the whole preferred to concentrate on the more lurid stories of randies and strikes, riots and brawls. There was, it has to be said, no great shortage of material.

CHAPTER TEN

Truck and Trouble

Some serious riots took place on Wednesday and Thursday, between the English and Irish labourers engaged on the North Midland Railway at Rotherham, the former attacking the latter in a large body with bludgeons and driving them off the line, for working at 9s. a-day lower wages. Military and even artillery were called in, thirty ringleaders secured, and four committed. The Riot Act was read, and it required all the energies of the local magistrates to restore peace, the whole county being in a state of fermentation. No lives were lost, but several were wounded, and some swam the river at Green Area to escape.

Felix Farley's Bristol Journal, 20 October 1838

The navvies who crowded into camps, shanties and lodgings all over Britain were a mixed lot, but inevitably they formed themselves into self-contained groups. The division might be based on nationality, or might simply consist of local men versus the rest. One result of this separation was that antagonisms sprang up between the different groups. The most obvious cause of any trouble was the resentment felt at 'strangers' moving into an area and taking over work which might otherwise have gone to local people. The fact that the strangers were often more experienced at the work made little or no difference: prejudice feeds on itself. Ugly scenes could occur anywhere. In 1845, trouble sprang up between the Irish and the native Welsh workers on the Chester and Holyhead Railway. 'On Friday a body of men from Aber, chiefly brick-makers & I believe Welsh, came to Bangor & attacked some Irish who are in Jackson's employ and drove them off the works.'[1] The military had to be called in. The company, however, seemed far less concerned about the fighting and the Irish navvies' loss of work than it was about the question of who was to pay the cost of quelling the riot. In the event the railway was left with a bill for £30 for the troops.

Incidents between different nationalities were all too common, and the flames of riot were most often fed by unsubstantiated rumours that one group – as often as not the Irish – was working for low pay. But the same arguments could flare up just as ferociously between much more local groups. This very full report from the Bristol area gives a clear idea of how disagreements could quickly turn to violence:

> On Monday last a number of navigators working on the Great Western Railway, amounting to upwards of 300, principally natives of the County of Gloucester, tumultuously assembled, and made an attack on the workmen employed at tunnel No. 3. Keynsham, who are most of them from Devonshire, and the lower parts of Somerset. The ostensible motive for the attack was a belief that the latter were working under price; to this was added a local or county feud, as the rallying cry of onslaught was 'Gloucester against Devon'. The result was a regular fight with various dangerous weapons, ready at hand, such as spades, pickaxes, crowbars, &c – The contest was long and severe, in which several were most dangerously hurt, & one man was obliged to be taken to the Infirmary, but no one was killed. The insubordination continued for several succeeding days, and was not repressed without the aid of the military.
>
> From further enquiries we learn, that a jealousy has long subsisted between the Devon men, & those from other counties, but the immediate cause of the late outbreak arose from the following circumstances. A few weeks ago, the company arranged with the contractor to take the working of the tunnel No. 3 out of his hands, and they then engaged a number of 'gangmen' who were paid a stipulated sum per yard for the work done by their men. It appears that the Devon men have been long accustomed to the use of a ponderous instrument called 'a jumper', for breaking the rock, and that they were thus enabled

to execute more work, and accordingly gain higher wages, than the workmen from other neighbourhoods; and the 'Gloucester men' being at least three times more numerous than the Devonians, the former came to a resolution of driving 'the strangers' from the works. The consequence has been that during the past week very few men have been employed on the Western railroad in this neighbourhood. In the several affrays there have been many heads broken, but we have heard of only one serious injury, and that was upon a man named Richard Thomas, now in the Infirmary, who it is feared has received a severe injury of the spine.

It would seem by the fact of a similar riot having taken place at the same time on the London end of the line, that there was some secret understanding and concert among the discontented workmen. The following is from a London Paper:-'We regret to state that a most desperate and alarming affray took place on the evening of Sunday last, betwixt the English and Irish labourers employed upon the GWR. It was again renewed on Monday and many of the results are anticipated to be fatal. The riot is understood to have arisen in consequence of the Irish party having proposed to work for lower wages than their English fellow-labourers. The atrocities upon both sides have been of the most brutal and unmanly description, and but for the interference of the local authorities aided by a squadron of the 12th Lancers, the most lamentable consequences must have ensued. Twenty-four of the rioters have been committed to Clerkenwell Prison, where they will now remain until a further examination.'[2]

Such riots were serious enough matters in all conscience, but at least one can see a certain logic, however perverted, at work. They were riots that reflected real concerns – wage rates, keeping local jobs for local people, protecting territory. And, it has to be admitted, they were sometimes effective – the strangers were driven away and wages did climb back to their old levels. But serious riots were not limited to the railways, navvy against navvy. The nineteenth century was not a time of peace and tranquillity. The Chartist upheavals that culminated in the Peterloo massacre, the 'Swing Riots' in the countryside, and the machine-breaking of the industrial regions were simply extreme examples of disturbances that welled up in Britain, mirroring on a smaller scale the giant upheavals of revolution that tore through Europe. Widespread discontent found relief in sporadic bursts of violence, and the presence of large numbers of rough, tough, itinerant workers in a region was in itself enough to cause trouble. One of the most serious outbreaks occurred at Preston in May 1838 and involved men from the North Union Railway and local weavers. It all began mildly enough when two Irish navvies asked for credit at a local shop. They were refused. Convinced that one of their fellow navvies had turned the shopkeeper against them, they set off to hunt him out and take their revenge. They decided, rightly or wrongly, that their man was being hidden in a local cottage, and when they failed to find him they smashed the place up and left, frustrated and in a vile temper, for the local pub. Here chance intervened with an unhappy combination of circumstances. The navvies had just been paid and would probably have had the place to themselves, but the local mill was closed due to an accident at the works, and a party of weavers with no work to go to were also drinking in the 'Sumpter House'. By now the two Irishmen's anger had turned from the man who had first caused offence to the locals who had 'hidden him'. A quarrel soon developed and the Irish told the weavers to get out or prepare to fight. The weavers left the pub but came back armed with cudgels and gave the Irish something of a thrashing. Now the trouble really started. The Irish, in turn, went back to the camp to recruit reinforcements. Soon they were marching into town, attacking anyone they met along the way. A couple of innocent carters were so badly beaten that, as the local paper put it, the hope of their recovery was 'exceedingly doubtful'.

The next day, the magistrates ordered the two parties to separate and return to their homes. They did so, but it was only a temporary reprieve, a pause to marshal their resources. The press reported the outcome:

> Both parties determined to fight it out, and accordingly they armed themselves with guns, pistols, pikes, knives, and other weapons, and assembling to the number of about eight hundred

> near the house of a man named Smith, had a regular fight. One man named John Trafford was shot through the body with slugs, and after walking a few yards fell down and expired. Several of the Englishmen received dangerous gunshot wounds. Two Irishmen are at present in the Preston Infirmary, one named Cassidy, and the other Kavanagh. One of them has had both his arms broken by a gunshot. A man named Bacondale received three gunshot wounds, one of which (through his loins) it is believed has entered his bladder. He is past the hope of recovery. A man named William Robinson received a gunshot wound in the arm and had his skull fractured. He is not expected to recover. The Irishmen deny having carried fire-arms, but it is rather a curious circumstance that six or eight Englishmen should have been shot if the opposite party had not carried fire-arms. It is believed that upwards of twenty Irishmen have been severely, if not dangerously wounded, but few of the belligerent parties dare come forward to give their evidence in a proper manner. These doings having been made known to the magistrates, a party of the 86th Regiment of foot were sent for from Blackburn, and they arrived after all the mischief was over.[3]

Such stories were of more than local interest. They were reported in national and local papers throughout the country, adding to the navvy's reputation as a man at best only one stage removed from a wild beast. It is easy when reading such accounts to see them as typical of navvy life, but one has to remember that they received such massive coverage precisely because the stories were lurid, sensational and, literally, extraordinary. No newspaper wanted an item that read: 'Several hundred navvies take day off – nothing happens.' There were real problems, all too many of them, but there were also works which went ahead peaceably, with virtually no bother at all. When the authorities wrote to the Taff Vale Railway demanding that they pay for a police force, the engineer was quite indignant: 'There is no immediate necessity for this measure as our men are *quite quiet*.' In his view the local authorities wanted to get the company to pay for police in the area rather than have to pay for them themselves.[4]

Once railway contractors started in an area, the navvies proved convenient scapegoats for anything that went wrong. When there was an outbreak of sheep stealing near Bramhope tunnel it was inevitable that suspicion should fall on the railway workers, but the local police inspector proved to be a veritable Sherlock Holmes:

> I know a gang of disreputable characters at Horsforth and from traces of footsteps, the length of footsteps, the way in which the sheep are killed, and other circumstances, I am satisfied the Horsforth thieves are the parties committing these depredations, and not your workpeople.[5]

He trusted the evidence of the footprints, set a watch on the Horsforth men and they were caught red-handed.

Yet railway navvies were no more virtuous innocents than they were dyed-in-the-wool villains. They were not gentle, retiring souls, and the anonymous army that moved around the countryside provided cover for more than one criminal. Nor, in spite of protestations of moral outrage, were their employers models of pacific virtue. Brunel's army that marched on Mickleton tunnel was acting well outside the law, and when Peto stood for Parliament he did nothing to prevent his navvies from intimidating the electorate into supporting him, even if he did not encourage them. For many years, and on many lines, railways were built under a system which seemed designed to encourage men into outbursts of drunkenness that were always liable to end in violence.

The greatest problems occurred when men were paid, particularly where payments were made in beer houses, and when there were long gaps between pay-days. Some who spoke out against the system seemed to suggest that the fault lay with paying the men at all. Thomas Beggs of the Scottish Temperance League declared:

> All my observations of the habits of the working classes, and the influences operating upon them would tell me this, that it is always bad policy to let working men, particularly the lower classes of operatives, have large sums of money in their possession at one time.[6]

Even when contractors paid out regularly, they generally gave the money to the gangers who, as one observer noted, were usually men whose job depended more on the strength of a good right arm and a big fist than on any other virtues.

> The ganger merely appears to have been selected from his hardy conduct, and his power to keep his men in order. I have in many cases had in evidence the fact that he is a person of very violent character, and not fit to have labourers employed under him.

Arguments over pay were settled by a thump on the head and then sealed with a few – or more likely a great many – beers. Once started, the drinking might go on for days, particularly when, as on the Caledonian line, payments were only made monthly. Then the navvies might stay drunk and quarrelsome for five days at a time. Commentators were inclined to blame the Irish, but really things were just the same on the London and Birmingham, where a decision was taken not to employ Irish workers at all, as Robert Rawlinson, one of the engineers, explained when questioned by the 1846 Committee.

> Did you find the men manageable and orderly? – I always found the men I had to do with, amenable to the instructions I had given them. As a matter of course, I took care not to go amongst them, if there was drunkenness or rioting, and they were in crowds; it would not have been safe or judicious to do so. Upon this work, I had no reason to complain, and no man ever resisted my authority.
>
> Did they lose any time after the pay-day? – Very frequently; in Northamptonshire, after Midsummer, there is a custom there, all the villages have there annually feasts . . . and so it would go on for three months, and nothing would keep these men from the feasts, and joining in the revelry and drunkenness.

Many of the problems and arguments started because of the iniquities of the 'truck' system. Navvies were often paid not in cash but in tokens or tickets that could be exchanged at the tommy shop. There were justifications made for the system. A navvy would arrive penniless at the workings and perhaps have to find the means with which to live for up to a month before pay-day arrived. In these circumstances, the truck system was used in effect to provide loans. If cash had been handed over there was always the chance that a man might disappear before he had earned it, but with a ticket only valid at the one place, the navvy was tied to that spot, for if he left all he had was a worthless piece of paper. The other argument advanced was that the system enabled a contractor to supply the necessities of life at a remote location and, as one sub-contractor openly admitted, it was also a way of solving cash crises. He did not always have enough money to meet wage bills, but at least no one starved.

Many of the arguments were spurious. Although outsiders and superiors seldom looked on navvies with a very kindly eye, most agreed with one of the engineers at Woodhead tunnel, who stated that 'it would be a disgrace if they saw a companion or a comrade without his dinner; they would divide what they had got with him'. The problem of waiting for pay was easily overcome by making weekly payments, as many of the better contractors did. And it was, after all, scarcely the navvies' fault if a contractor was short of cash. On the other hand, the opportunities for abuse were all too obvious. If a man was tied to getting all his supplies from just one place, then he had no option but to pay the price asked at that place – no use his pointing out that bread was twice the price of that in the shop down the road if he had no means of going there. Even worse troubles occurred when debts were run up over many weeks. At the Woodhead tunnel pay-days could be as much as nine weeks apart, and no navvy would have had the least idea whether the beer bill with which he was presented at the end of that time was accurate or not – he might well have had trouble remembering what he had had the previous night, let alone two months ago.[7] One contractor took the trouble to refute the allegations, claiming that he provided food at fair prices and refused to sell beer.[8] But there can be no doubt that the system was hated by the men and equally disliked by many of those who employed them. The engineers, in particular, took a dim view of contractors or gangers selling beer. For the contractor, it was a case of 'heads I win, tails you lose': if the men worked they earned him money, if they went off drinking he made a profit on the beer. The engineers saw beer being sold along the line and knew that if there was a stoppage for some reason, maybe for an hour or two, it was very difficult to get the men back to work. Thomas Jackson, the contractor, explained what all too often happened:

> If it comes on wet after 10 o'clock in the morning, the men knock off, and if there is a beer-shop in the neighbourhood kept by a ganger, and if they have not money they know they have a claim upon him for the previous two or three hours they have worked; and as long as they have a claim, one or two, or three or four, of them, will insist on getting beer to the amount of what they have done.

The truck system was in general use on the Trent Valley Railway. A local gentleman, Sir Charles Wolseley, wrote in high indignation to the company of which he was himself an important shareholder:

> There is shameful work going on on our line of railway, particularly on the score of the Truck System. Now you know as well as I can tell you that cannot be carried on without the aid of those who are subordinate agents – will you then be kind enough to inform me to whom I am to apply to put a stop to it if possible – There has been here one strike already and I know from scores of the working men that they are naturally dissatisfied – there are also a strange set of fellows employed as *time-keepers*.[9]

The local magistrates declared that the only people they could act against were the gangers, by distraining their property – but as they were outsiders, living in lodgings, there was no property to be seized. They thought that Parliament should legislate for weekly pay, ban the truck system and make the railway company responsible for ensuring wages were paid – and paid when the money was due. Until then they could only call on the company to see what could be done and lay out the grievances of the men. The company responded, but in its response there seems rather more emphasis on keeping the men in order than in preventing the abuses which had made them disorderly in the first place:

> I have made inquiries into the complaints made by the magistrates of Brinklow, and find that the statements in their letter are true, that the men were on one occasion paid after a lapse of five weeks and that the men were disorderly on that occasion. On the recommendation of the magistrates the contractors have appointed four policemen to be stationed between Rugby and Brinklow, and they are to be under the direction and control of the county police from the district.
>
> The men will for the future be paid once in four weeks and in the intermediate time they will be allowed some money at the rate of two shillings per day on account. The sub-contractors keep provision shops for supplying the men, and invariably give the men tickets for the provisions they require. This causes much dissatisfaction among the men as they are compelled to go to a particular shop, where they are charged from twenty to fifty per cent more for most articles, than they would be at a common shop.

Occasionally the men were more fortunate. Workers in East Anglia were paid in the usually noisy, crowded room and hustled away as soon as the money was in their hands, with no time to check it. One man who should have had £2 9s. found he was a pound short and eventually took his case to the magistrates. It was the ganger's word against that of the workman, and on this occasion, at least, the workman was believed – but even the magistrate seemed surprised by his own decision.

The truck system was roundly condemned and laws were passed forbidding it, but it went on regardless. Sub-contractors found they could just about keep going with the help of the 'Tom-and-Jerry', as the local beer shop was known, while some greedy men made more from truck than they did from the works, running tommy shops where a sovereign bought fifteen shillings worth of goods. Even when abuses were blatant, nothing was done. A magistrate from Atherstone, asked to help the men get the pay they were owed, simply replied that the law did not apply when it came to arguments between employers and their men. If a magistrate felt that was the case, then what were the men to think? If that was the official view, then they were quite literally 'outlaws', at least as far as their own grievances were concerned. Contractors signed documents which spelled out in detail that they must not run tommy shops, nor supply beer, but the company was as reluctant as the magistrate to step in on behalf of the men in opposition to the contractors. It preferred to stay aloof in many cases. So the navvy was inclined to take matters into his own hands.

What is noticeable in looking through the records is that while there are numerous accounts of riots, there are comparatively few mentions of strikes, despite the fact that there was far more likelihood of persuading a contractor or a company to pay more wages, abandon truck or agree to regular payments by hitting the employer in his pocket rather than another workman on his head. The reason for the lack of strikes is not hard to find. The organization of the work between individual contractors and sub-contractors meant that it was almost impossible to get the different gangs to maintain any long drawn-out campaign. The interests of the big contractors seldom coincided with those of the small; the sub-contractors worked almost on a day-to-day basis and they knew that, given a prolonged stoppage, the gangs would soon begin to drift away to find work somewhere else. Even when they did get together, the employers could call in the authorities.

> On Friday the whole works of the Scottish Central line from Peterhead to Thornhill amounting upwards of 3000 sturdy labourers armed with shovels assembled in front of the offices at Stirling when they demanded an advance of wages of 3s per week. The manager of the works considering the shortness of the day the rate of wages at present being 2s 6d per day did not think it proper to comply with what seemed to be an unreasonable demand; and as it was feared that some disturbance might take place every precautionary measure was adopted and the workmen are now regretting their rash step and are resolved to return to their work. Some of the ring leaders have been paid off.[10]

The statement that the men were now 'regretting their rash step' is not as ominous as it might sound. The men had, in effect, tried for a better wage, seen that they were not going to get it and accepted the situation. Even the 'ringleaders', usually just the unfortunates who had volunteered – or been volunteered – as spokesmen, could generally rely on finding more work elsewhere. The employers also had another potent weapon in their armoury: if one set of men stopped work, another set could always be found to take their place.

> The labourers on the Richmond railway 700 in number have struck for an advance in their wages of 6d a day. The contractor refuses to give it; declares that he will not again employ one of the strikers, and has advertised for other hands promising '3s 6d and 4s a-day and no tommy-shops'.[11]

If strikes were rare, then successful strikes were rarer still, and the old pattern of steady work, interrupted by sudden outbursts of drunkenness and wild behaviour, persisted.

Pay-days invariably ended in a 'randy', a navvy name for what might be no more than a monumental booze-up, but which one engineer described as an occasion for 'every sort of abomination, lewdness and bad women'. Lurid pictures were painted of pay-days: 'Like dogs released from a week's confinement, they ran about and did not know what to do with themselves . . . Their presence spread like a plague.'[12] Generally, however, they were self-contained, restricted to navvy camps or the pubs which they would more or less take over and make their own. But there was always the possibility that a small incident could blow up into something altogether more unpleasant, dangerous and even lethal. And often, it seems, it was the sense of resentment against the outside world, the sense of having to look after their own, that fuelled a quarrel and turned it into a full-scale riot. This is what happened on just such an occasion as described by Graham Spiers, the Sheriff of Edinburgh.

> It was upon the evening of a pay-night. Some of the police on the railway had apprehended two Irish labourers upon a trivial

> charge, the suspicion of stealing a watch, and they were lodged in a temporary lock-up house near the line; their companions immediately proceeded to the huts in the neighbourhood where a great body of labourers were, and having obtained their assistance, they returned to the prison where those two labourers who had been apprehended were confined. They broke it open and overpowered the policemen, who were defending it, and rescued their companions. After they had done so, and within a very short distance of this lock-up, they met two other policemen coming towards the lock-up house, not aware of what had taken place; they attacked them, and killed one of them and hurt the other. This outrage was committed almost entirely, I may say, by Irish labourers. It happened upon a Saturday night . . . and upon the Monday the Scotch labourers assembled in a large body, and under the pretext of avenging the death of this policeman, who was a Scotchman, they proceeded to the huts of the Irish people, which they burned, drove the Irish away, and committed various other acts of outrage.

If the law was not to intervene between men and employers, if the company regarded arrangements made by the contractors as not being any of its concern, then what was to be done to stop what were undoubtedly serious outbreaks that terrorized a neighbourhood and led to injuries and deaths? One possible answer was to transform the navvies into sober, moral citizens. What was needed was:

> a Clergyman – energetic – zealous – and judicious, a gentleman easy of address, free of access, possessed of skill and tact – and withal a plain and earnest speaker, and devoted to his work.[13]

The writer quoted the good example set by the Great Western Railway and enclosed a press cutting to prove the point. He did not, however, mention that on the same line men were being fined 5 shillings each – roughly two days' wages – for working on the Sabbath, though nothing was done about those who ordered them to do so. The men on the Great Western had more spare time on the Sabbath than perhaps they would have wished. Nevertheless, many workers did seem to appreciate the fact that the clergy were taking a friendly interest in their welfare.

> On Sunday last the Rev. F. Close preached in a neat wooden temple near the Birmingham depot at the extreme end of the Queen's Road, Cheltenham, adjoining the Gloucester road; the object of which was to induce the men employed on these railway lines to attend and hear the word of God on Sunday afternoons. On this occasion several hundred persons were present chiefly of the class for the benefit of whom the building was designed. The rev. gentleman having read the usual service of the church preached from Acts XVI.29 and two following verses and was listened to with the most marked attention. The rev. gentleman in conclusion stated that it was his intention to preach every Sunday afternoon in this little building now dedicated to God and he devoutly hoped that some good would result from the service. Mr. Oldham, the contractor, then stated that as the building did not seem large enough he would be most happy to make it double the size by the following Sunday. We need not add that the offer was most readily accepted.[14]

A number of companies agreed to pay – or at least share – the expenses of having a clergyman work among the navvies. Some made evangelism part of a more general package of benefits, though as Captain Moorsom of the Chester and Holyhead Railway clearly explained in a letter to the Bishop of Chester, it was not the welfare of the navvies that was uppermost in his mind:

> The Directors knowing from experience the delay in the progress of the Works, the damage inflicted on adjacent properties & the annoyance experienced by the resident gentry, as well as the loss accruing to the Contractor from the loose unsettled habits of, & constant changes in the description of men engaged in Railway Works, are desirous of holding out inducements to uniform and steady industry & to promote the growth of moral habits.[15]

Captain Moorsom therefore proposed banning truck, insisting on regular payments and appointing two scripture readers. The clergy were generally zealous men who, realizing that the men might not come to them, went to see them where they lived and worked. There was among many of the navvies a genuine desire to be a part of civilized society. William Breakey of the Town Missionary Society went to the Chester and Holyhead Railway where some 700 Welshmen were at work. He sold 374 Bibles, printed with Welsh on one page and English on the other. Another clergyman, however, who gave away testaments, had the unhappy experience of seeing the navvies in the nearest town selling off their texts for beer money.

The clergy had at best a very limited effect in bringing religion, sobriety and high moral standards to the navvies. This was in part due to their background: a university education is more easily deployed from the pulpit of a parish church than in a wooden hut in a shantytown. The men were more inclined to listen to preachers such as Peter Thompson, known as Happy Peter, who had himself been a navvy.

Among the contractors, Peto, whose second wife was a Baptist and filled with evangelical zeal, was famous for the efforts he made to bring religion to his men. He employed lay preachers and scripture readers, handed out improving texts and, once he was assured the request was genuine, handed out Bibles. All these were notable efforts by individuals, but it was not until the 1870s that an organization was set up to preach the gospel – and sobriety – to the navvies. Mrs Elizabeth Garrett always said that she was influenced by a ceremony in Otley churchyard at the memorial to the twenty-three men who died on the nearby Bramhope tunnel (the memorial still stands, a small-scale replica of the tunnel itself). Mrs Garrett was later to meet a Leeds vicar, Lewis Moule, who founded the Navvy Mission Society, and she in turn founded the Christian Excavators' Union. It was not a great success at first – after a decade there were less than 300 members out of an estimated navvy population of 70,000. Over the years the Society may not have made many converts but it did do a great deal in a practical way – supplying reading rooms and entertainments, setting up soup kitchens in hard times, and tirelessly sending out its lively propaganda and newsletters. Although only a small proportion of navvies could read, the reading rooms were full, for the literate read out passages to the others. It seems quite likely that the audience was more interested in the news stories, often lurid accounts of accidents and swindles, than in the purple prose of evangelism. The men had their own, if not entirely convincing, song to sing.

> Yes, I am an English navvy: but, oh, not an English sot.
> I have run my pick through alcohol, in bottle, glass or pot;
> And with the spade of abstinence, and all the power I can,
> I am spreading out a better road for every working man.

It is doubtful if many English navvies sang those words with any degree of conviction.

It is easy to point to a certain hypocrisy in the evangelical movement which put such emphasis on spiritual welfare, when there was a pressing need to do something about the physical and practical difficulties that beset the navvy life. But members of the movement did go among the men, spoke to them directly and showed a genuine concern for their welfare. The Reverend Robert Thompson, chaplain to the workmen on the South Devon, took a positive role when the workmen went on strike to protest against monthly payments. He acted as arbitrator and managed to get payments made fortnightly, which cut down on the use of the tommy shop. Not that the Reverend had much of an opinion of the workers. He thought local shopkeepers would be 'silly' to give credit to the men: 'There is not an atom's worth of honour among them.' This was the harsh judgement of a man prepared to help the navvies. Others simply regarded them as little better than wild beasts, existing quite outside the bounds of decent society. Authority was ready to raise an outcry against the rowdy, bawdy and sometimes riotous life of the navvies and to punish the wrongdoers. Less enthusiasm was shown for remedying the conditions from which such behaviour sprang. Navvy life was hard and for many it was the only life they would ever know. Among those who came forward to give evidence to the 1846 Committee was Richard Peace who had set out on his itinerant life thirty-two years before, as a 20-year-old navvy on the Lancaster Canal. Another was Thomas Easton, twenty-seven years a navvy, who probably spoke for most of them when he was asked if he had known ups and downs: 'Yes, many up; not very far up; but many down.'

CHAPTER ELEVEN

Men at Work

He will hear the blasting of the stone in the tunnel beneath him, or his attention may be caught by the noise of the horses and men working at the mouth of the shaft; confusion will appear to reign wherever he looks.

Account of tunnel works on the Great Western Railway

Work changed surprisingly little over the whole of the great age of railway construction. Mechanical excavators were in use in America as early as 1842, but the British obstinately refused to consider them: they may have been all right for overseas, but they were not suitable for Britain. It was not until the building of the Great Central, the last main line, completed in 1899, that steam excavators, forty of them, were introduced in any major way. It was a last grand gesture on a line which was, alas, a monumental folly, an act of megalomania on the part of its promoters. Until then the work depended, as it had from the beginning, on muscle power. Many lines never even took the obvious step of having contractors or the company supply locomotives to run the trucks of spoil that were produced every day. The track was laid, and along it the horses plodded. If steam appeared at all it was in the form of the stationary pumping engine, an essential part of all major tunnelling works. The railways were the wonders of the nineteenth century, but they were built using the technology of the eighteenth century.

It is difficult at this distance to imagine the scene at a railway construction site – the earth torn apart to make way for the line; grass and trees and shrubs removed to lay open a scar of bare earth and rock that stretched across the land, temporary rails and tracks, the hiss of steam engines, the rattle of carts and barrows, and everywhere men, hundreds, thousands of them spread out all along the line. In the country, at least, the course of the works was clear, but in towns it was an inconceivable jumble. Dickens described the scene in a memorable passage:

> Houses were knocked down; streets broken through and stopped; deep pits and trenches dug in the ground; enormous heaps of earth and clay thrown up; buildings that were undermined and shaking, propped by great beams of wood. Here, a chaos of carts, overthrown and jumbled together, lay topsy-turvy at the bottom of a steep, unnatural hill; there, confused treasures of iron soaked and rusted in something that had accidentally become a pond. Everywhere were bridges that led nowhere; thoroughfares that were wholly impassable; babel towers of chimneys, wanting half their height; temporary wooden houses and enclosures, in the most unlikely situations; carcases of ragged tenements, and fragments of unfinished walls and arches, and piles of scaffolding, and wildernesses of bricks, and giant forms of cranes, and tripods straddling above nothing. There were a hundred thousand shapes and substances of incompleteness, wildly mingled out of their places, upside down, burrowing in the earth, aspiring in the air, mouldering in the water, and unintelligible as any dream. Hot springs and fiery eruptions, the usual attendants upon earthquakes, lent their contributions of confusion to the scene.[1]

Looking now at the urban scene, railways seem so familiar a part of the landscape that one cannot imagine what an upheaval their building caused. The London and Greenwich swept into London on a viaduct that was nearly four miles long, and for months the surrounding streets were full of carts, delivering bricks and timber, and men at work, while the scaffolding rose above the rooftops.

Railways were, it seemed, unstoppable - not even the dead were allowed to get in their way. The line from York to Leeds passed through the edge of the parish church graveyard: the tombstones that were uprooted were laid out on the new embankment and there they remain. A route was blasted through to Liverpool, creating the rocky chasm of Olive Mount cutting, which attracted throngs of sightseers during its construction. Here not just a few gravestones, but an entire church had to be shifted out of the way, and moved again during later widening operations. By the end, the church had moved three times. In some cities the railways skulk out of sight as though the city fathers were ashamed of such a vulgar eruption in their midst - Edinburgh is a classic example. Such delicacy extended to spa towns such as Harrogate, where the builders were forced to keep the lines out of sight in an otherwise unnecessary cutting so as not to spoil the view. In other places, the lines march triumphantly forward, as in what must be one of the grandest of all city entrances - the approach to Newcastle from the south over Robert Stephenson's bridge over the Tyne. Clearly, town and country presented engineers with quite different problems and produced quite different construction scenes, but the fundamental work remained the same. Railways could be built on the flat, which generally presented few problems; they could pierce hills by cuttings or tunnels and cross valleys on embankments, bridges and viaducts - all of which presented very real and very different problems. The remark that flat land presented few problems, however, needs to be qualified - one should say, provided the ground was firm. George Stephenson's Herculean efforts to conquer Chat Moss are well known, and we have already met the 'quaking bogs' of East Anglia, but all of these seem almost insignificant when set against the problems faced by the engineers and men building the East Lancashire line at Haslingden. The ground was just about as bad as it could be, an unholy alliance of peat, sand and gravel, all well and truly saturated with water. The peat alone was estimated as being 20 ft. deep. The obvious answer was to pile on spoil until the ground stabilized. Every day for three months trains consisting of trucks piled with earth were hauled in and their contents tipped into the morass, which accepted the offer with a gurgle as the earth sank from sight.

Only at the end of that time did even the first signs appear that the process of stabilization had finally begun.

The great deep cuttings are known to us from the work of the artist J.C. Bourne who took his pencil and sketch-pad to record the men at work at Tring and Blisworth.[2] These are dramatic scenes which, perhaps more than any other, give a sense of the sheer scale of the enterprise. The work began modestly enough with the digging of a 'gullett', a cutting just wide enough for a track to be laid for the wagons that would take away the excavated spoil. Mostly the wagons were hauled by horses, with a brakeman in charge who relied on a simple lever on the last wagon to stop his laden train. It sounds, at first, as if it might have been a rather happier job than actually digging at the cutting, but not so according to contemporaries. His duty was 'anything but pleasant for, what with the roughness of the roads, and the action of the springless vehicles on which he rides, the shaking he receives in his journey seems sufficient to reduce every joint in his body to a most unsatisfactory condition of laxity'.[3]

The actual work of cutting depended on the nature of the ground. In rock or chalk, common black powder was used to blast a way through. In soil and clay, the work fell to the navvies with spades and pickaxes. The men would work at vertical faces, up to 12 ft. high. The technique was to cut channels down the sides of the face and undermine it, leaving solid pillars of earth as supports. Then, when everything was ready, the supporting pillars were cut away and the whole section levered out. Ideally it would topple as a piece; otherwise it would simply subside in a heap. The bigger the face that could be removed at a time, the less work was required, and the greater the profit. Equally, the bigger the face the more likely it was to collapse of its own accord, crumbling down on the men working below. It was a practice which depended on skill, experience and judgement. When these were lacking, the result could be injury and death. An engineer, Rawlinson, described how the system worked as he had seen it. When the face started to move, a man posted at the top of it shouted a warning and the men beneath scattered. If it failed to move when the pillars were removed, bars and rods were pushed in at the top to lever the mass away. He remarked that sometimes the men were inexperienced, or the winter evening made seeing difficult, and the face crashed down, killing and maiming. In such a case it was, as far as he was concerned, clearly the men's own fault.[4] Some cuttings had to be made in atrocious conditions. The famous London clay of which Robert Stephenson had been warned lived up to its reputation. After weeks of work on a cutting on the London and Birmingham line into Euston, a period of prolonged rain set the clay sliding and, before anything could be done, the cutting so laboriously executed was filled up again. On another cutting where clay was encountered it seemed for a time that the work would never be done – as fast as the clay was dug out, it flowed back in again 'like porridge'. It must have been pure misery for the men fighting this cold, slimy, cloying mass.

Once the faces had been toppled, the debris had to be cleared and loaded into trucks. This was back-breaking work and it was here that the experienced navvy scored over his inexperienced neighbour. On the Chester and Holyhead line, five Welshmen were needed to fill each truck in the early days, but only three English navvies. The best of the gangs could do even better. Thomas Brassey in the early years employed former canal navvies as far as was practical.[5] He reckoned that in a full day's work a gang could fill fourteen sets or trains of wagons, working two men to a wagon. Each wagon held 2¼ cubic yards of spoil, which had to be lifted to a height of 6 ft. to clear the sides of the truck. So each man lifted 16 cubic yards of earth and stone over his head. Such numbers often seem meaningless, but anyone who has ever done any gardening might like to think of a hole 3 ft. deep, 3 ft. across and 15 yards long, then imagine the effort involved in shovelling out all the earth and clay and chucking it over a head-high wall in one day, and then going out and doing the same thing the next day, and the next, and keeping going for weeks on end. That was what the navvy did, and some did even more. If there were a hundred men at work, there would be three trains of wagons in use; one being filled, one out on the rails and the third being emptied. It was a continuous process which allowed for no respite, as demanding as a modern production line.

Where the spoil was not being carried away for building an embankment, it was taken out by the shortest route – straight up the side of the cutting, and in deep cuttings this involved the barrow runs. At Blisworth there was a lift of 10–12 ft. of an unappetizing mixture of gravel, clay and marl resting on a bed of limestone. The material was blasted – 100,000 pounds of gunpowder was reckoned to be used – and loosened with pickaxes, then loaded and taken to the planks laid up along the side of the cutting. Here the barrow was hitched on to a rope, which ran over pulleys at the top of the bank to a horse. At a signal, the horse was led forward and the barrow with the man to steady it began to rise up the slippery planks. It sounds simple in theory but was less so in reality.

> The practice of running, though common, is dangerous, for the man rather hangs to than supports the barrow, which is at once rendered unmanageable by any irregularity in the motion of the horse. If he finds himself unable to control it, he endeavours, by a sudden jerk, to raise himself erect; then throwing the barrow over one side of the board, or 'run' he swings himself round and runs back to the bottom. Should both fall on the same side, there is great risk of the barrow with its contents falling on him before he can escape.[6]

And if going up was bad, coming down again was no great improvement, as the man now led the way, galloping down with his barrow trundling behind him. It was a skill which had to be learned and not everyone managed it. Rawlinson, who had described the work in the cutting, also described the runs. A new man would set off, then halfway up his confidence would go, and when that happened he inevitably stopped and fell. Again Rawlinson was quite clear – it was the man's own fault, even if the experienced men had been deliberately piling on the work.

> There is a system called 'harrying' the men. If it is barrow work, for instance, they will so overwork him, that if he is not accustomed to it he will sink under it . . . if he is not a man of extraordinary muscular energy, he cannot stand the system.[7]

The material that came out of the excavations was not simply waste to be dumped. Most lines worked on the 'cut and fill' system –

material dug out from the hill was taken away to other parts of the work to build embankments. If there was good turf on the surface, this was stripped off and stored ready to be laid on the completed bank, while the main part of the spoil was taken straight from work on the cuttings to the next section of embankment. The quantities involved were prodigious: at Tring an estimated 1,400,000 cubic yards of chalk were removed – enough material for a bank 10 yards across, 10 yards high and 8 miles long. So the activity in the cutting was marked by equally frenzied work moving these vast quantities of spoil along the line. Accidents here were also commonplace, often caused by men taking unauthorized rides on the trucks. At Blisworth the trucks from the top of the cutting hurtled away down one incline where it was apparently by no means unknown for as many as twenty trucks to be smashed in a single pile-up if a train of wagons got out of control – anyone hitching a ride was likely to suffer a similar fate. But a far greater cause of accidents came when the trucks were run along the top of a partially made bank and tipped up at the end.

A common method of moving spoil involved the horse trip. Even when locomotives were employed for long hauls from cutting to bank, horses would still be used for the last part of the run. The engine would puff along, then the trucks would be uncoupled, points switched and the trucks trundled forward. The horse would be harnessed to an individual truck and would set off – often straining to climb uphill – and the essential was to get up a good trot to give momentum to the truck. Then, at the last minute, the horse was unhitched, pulled to one side and the truck sped on to crash against a barrier at the end of the line. The truck was designed to tilt forward so that the spoil was shot out to be spread by the waiting navvies. Again it was a matter of nice judgement on the part of the spreaders to be near enough to get to work quickly, but not too near so they were buried. The fillers had a special name for any truck they sent on that was particularly well filled – 'we'll run 'em in a red 'un'. The greatest risk was taken by the man, or more often the boy, who had to lead and unhitch the horse on the run.

A very young man named John Allen one of the drivers on that

> part of the Great Western Railway now constructing by the side of the canal on the road to Bathampton was on Thursday night brought to the Bath Hospital having met with a frightful accident. His foot slipped and he fell when a whole train of tram waggons went over his right arm and thigh completely smashing both limbs. The loss of blood was so great that he died soon after he had reached the hospital.[8]

By far the most dangerous part of all railway work was tunnelling, and here a whole new set of specialist skills were brought to bear, for the techniques and technology had developed in the long history of mining in Britain. And it was to miners that the engineers looked to find their sub-contractors and gangers. At the great Bramhope tunnel, when things were going badly, the engineer was able to announce that he had sacked the sub-contractor and relet 'to a set of 4 miners' who were making such good progress that they were already being offered premiums for finishing early.[9] This was also an area which not only required special expertise, but involved the company in a considerable outlay in equipment as well.

Work was generally begun by sinking shafts down to the level of the tunnel. These were generally circular, and some reached to great depths. At Woodhead, for example, the longest shaft was begun in the bleak uplands of Round Hill at a height of 450 metres (1,476 ft.) and was sunk down to the tunnel at the 310-metre (1,017-ft.) level. The techniques had scarcely changed for centuries: the work was still largely done by men with pickaxes digging ever deeper into the ground, and using wedges to break open rocks. The only modern aid was the use of powder for blasting more stubborn areas of rock. Sinking was laborious, but the more shafts there were, the more men could work in the tunnels themselves, for two groups of men could work out in opposite directions from the foot of each shaft. When completed, the shafts were generally lined with bricks, and many remained in use as ventilation shafts when the line was opened. These appear now as round brick structures on the tops of hills, and the progress of trains far underground can be marked by the line of smoke steaming up from the shaft, though these days it is diesel exhaust not coal smoke. At the top of the shaft there was a good deal of activity. First there had to be some means of getting men, materials and loads of spoil up and down the shaft, and the commonest device was the horse whim or gin (sometimes a steam engine). The horse plodded along a circular track, harnessed to one end of a long wooden arm which was in turn geared to the drum with its cable passing down the shaft. As the horse walked, so the drum turned and men and material were brought up or down. Few people outside the teams of working men ever glimpsed the underground tunnelling world, but those who did found it a dramatic experience:

> The descent by shaft No. 7, which is 136 feet deep is effected on a platform, without any railing or other security on the sides attached to a broad flat rope wound and unwound by a steam-engine, and is attended with no convenience (if the idea of a fall from giddiness, or from the breaking of the rope, be not allowed to intrude) except the hard bump with which your arrival at the bottom is announced to you. The works are being carried on each way from this shaft and which, every way you go, the same appearances meet you. The dark, dim vault filled with clouds of vapour is saved from utter and black darkness by the feeble light of candles which are stuck upon the sides of the excavation, and placed on trucks or other things used in carrying on the works; these which in your immediate neighbourhood emit a dull red light are seen gradually diminishing in size and effect, till they appear like small red dots and are then lost in the dark void. Taking a candle in your hand you pick your way through pools of water over the temporary rail among blocks of stone, and the huge chains attached to the machinery which every now and then impede your way happy and lucky if no impediment, unobserved in the dull, uncertain light should arrest your progress by causing you to measure your length on the wet and rugged floor. Pursuing your onward course examining by the way the appearance of the works and admiring the solid walls which nature has provided, you note every now and then a beautiful rill as clear as crystal issuing from some fissure in the rock trickling down the

sides of the tunnel, and helping to form one of the many pools and streams with which the floor yet abounds. Not during all this time have your ears been idle, the sounds of pick and the shovel, and the hammer have fallen upon them indistinctly; but as you advance they increase and the hum of distant voices is heard. The faint illumination before only just sufficient to make darkness visible, now becomes stronger and the lights which had been placed chiefly in line along the walls become more frequent, they dot the whole of the opening, being pretty thickly planted from the floor to the roof. The cause for this is soon apparent: as you advance, a busy scene opens before you; gangs of men are at work at all sides, and the tunnel which to this point had been cut to its full dimension suddenly contracts; you leave the level of the floor and scrambling up among the workmen stopping sometimes on the solid rock at others on loose fragments you wind your way slowly and with difficulty. Having been informed that a shot is about to be fired at the further extremity you stop to listen and to judge of its effects. The match is applied, the explosion follows and a concussion such as probably you never felt before takes place; the solid rock appears to shake and the reverberation of the sound and shock is sensibly and fearfully experienced; another and another follow; and with a slight stretch of the imagination you may fancy yourself in the middle of a thunder cloud with heaven's artillery booming around. You pursue your rugged path and having arrived at that part where the junction was made between the two cuttings, you have an opportunity of examining the roof and of admiring the solid bed of rock of which it is formed, and of appreciating the skill which enabled the engineer to keep a true course under all the difficulties of such a work. After traversing a considerable space within reach of the roof, you find your way to the bottom among a gang of labourers who are working from the other end and having arrived at the shaft at the Chippenham side of the tunnel you step upon the platform the word is given and you are once more elevated to the surface of the earth, and are glad to breathe the pure air

> and full of wonder at the skill enterprise and industry of your fellow-men.[10]

The beautiful 'rill as clear as crystal' was regarded less poetically by engineers, who found flooding to be by far the greatest problem they faced in the tunnels. Luckily for them, the nineteenth century had seen an enormous improvement in the power and efficiency of steam pumping engines, particularly in the deep wet mines of Cornwall and Devon. The ending of the Boulton and Watt monopoly which hampered all eighteenth-century development brought a new generation of powerful pumps, just as it gave men like Trevithick the chance to develop the locomotive. The engines, some small, some immense creatures in their own grand engine-houses, nodded day and night, working the pumps that kept the tunnel dry. Surrounding the shaft, with the whim, steam engine and pump rods, was often a confused scene, with carts and trucks taking away the spoil for dumping or banking, other carts bringing in coal for the insatiable engine or material for underground and, in the remoter tunnels, the huddle of shacks and huts that marked the workers' temporary home. But the important work was going on underground.

Blasting with gunpowder was for years the main means of moving the tunnel forwards, though towards the end of the century other more powerful explosives, such as dynamite, came into use. First a hole had to be drilled into the rock by hand, then the hole was filled with powder and rammed home or 'stemmed' with a metal rod. There was a good deal of controversy as to whether a more expensive copper stemmer would be safer than the common iron rod, which could produce sparks that ignited the powder. An assistant engineer at Woodhead said it would make no difference, but this report describes an incident at that tunnel:

> William Jackson, miner, No 5 shaft. He was looking over John Webb's shoulder, while he was stemming a hole charged with powder, when the blast went off, blowing the stemmer through Jackson's head, and killed him on the spot.[11]

In the old method of working, a thin rod was put in the hole and then packed round with clay, the rod was withdrawn and a powder trail laid. This was lit, and the man charged with the job scampered back to the shaft to be raised to safety. He did not always make it. Things improved with the introduction of safety fuses, but even then there could be trouble. One man had lit the fuse and the ascent had begun when the platform jammed in the shaft. With great presence of mind he signalled to be lowered not raised, jumped out and stifled the fuse before it reached the powder. The safety fuses were not always used, as another Woodhead engineer explained under questioning.

> In blasting in this tunnel was the safety fuse used? – No. Is that not more safe for blasting, than the common fuse? – Perhaps it is; but it is attended with such a loss of time and the difference is so very small, I would not recommend the loss of time for the sake of all the extra lives it would save.[12]

Even by Victorian *laissez-faire* standards this shows an extraordinary degree of cynicism.

As the miners advanced the faces, so the walls of the tunnel were shored up behind them with timbers – unless, as sometimes but very rarely happened, the rock was solid enough to stand unsupported. There was also a busy traffic of trucks of spoil being wheeled away from the face, as the debris of each explosion was cleared, to be raised back up the shaft. In this aspect, the workmen on Standedge tunnel between Manchester and Huddersfield had an advantage, for they worked alongside the old Huddersfield Canal tunnel. The two tunnels were interconnected so that spoil could be brought out by boat. This was of great value to the railway, but as those who remember the days when the canal tunnel was open, a great nuisance to later generations of boatmen. Each train passing through filled the neighbouring tunnel with its choking fumes. There were three railway tunnels in all at Standedge and each made use of the canal.

Behind the miners came the bricklayers to give the tunnel a permanent lining. Tunnelling was, in other words, an immense enterprise, far more than simply a case of digging a hole through a hill.

Some 'tunnels' indeed were worked in an altogether different manner. The Metropolitan Railway in London was made as a deep, open cutting which was then covered over. But by far the greatest tunnel of the nineteenth century was that under the River Severn, dug to provide a fast route on the main line to Cardiff.

In October 1879, when Sir Daniel Gooch of the Great Western was present at the opening of the bridge over the river at Sharpness, work on the tunnel was well under way and Gooch invited the guests to take a walk under the river: 'It will be rather wet, and you had better bring your umbrellas.'[13] This was to turn out to be one of the great engineering understatements of all time, for unknown to Gooch the Great Spring had just been excavated, a powerful underground stream that was even then in the process of flooding the workings and bringing everything to a halt. It was, it seemed, the final blow in what had been an unhappy phase in the construction of the tunnel. The plans were first put before Parliament in 1871 and work began two years later. There was a great surge of activity, with the first shaft on the Welsh side being sunk to a depth of 200 ft., and a small colony of workers grew up around it. But by 1877, headings had advanced less than a mile. This somewhat pathetic rate of progress was speeded up by sinking more shafts, using the islands under which the tunnel passed, but then in 1879 the catastrophe struck and the original workings were flooded. By now it was clear that the original contractors were wholly inadequate, and Thomas Walker took out a contract to finish the job.

The first priority was to bring in really powerful pumping engines, and that was done. The engines, built by one of the greatest enginebuilders of the age, Harveys of Hayle in Cornwall, were real monsters. The cylinders of these powerful beasts were enormous – one 70 in. in diameter, the other 75 in., and the larger of the two was able to raise 3,000 gallons a minute. But even these were unable to conquer the Great Spring – the inundation was so immense that it was invariably honoured with capital letters. It was obvious that the only solution was to seal off the torrent, and the work could only be done by divers. It seemed an impossible task, not only because of the intrinsic difficulties of shutting off the flow, but also because of the need to use the same divers to carry out underwater repairs on the pumps. It is hard to imagine the nightmare of these dark workings, but at last a diver was sent down with a real chance of restoring the damage, and sealing off the spring.

> He started on his perilous journey armed with only a short iron bar, and carefully groped his way in total darkness over the debris which strewed the bottom of the heading, past upturned skips, tools, and lumps of rock, which had been left in the panic of 1879.

To complete his task, he had to descend 30 ft. and then make his way for another 1,000 ft. along the heading. He almost made it, but in the end his own equipment defeated him – his life-line floated up to the roof of the flooded tunnel and dragged along the roof until it was quite simply impossible to drag the tubes any further. A last desperate attempt was made using what was called 'a knapsack' of compressed air. The diver faced terrifying hazards. Moving in the dark, his vital air supply could catch on any obstacle, or one of the flexible tubes could snag and rupture, in which case he would, quite simply, die. But he made it. He repaired the pumps, set the valves and returned. It had been an heroic expedition, using experimental and untested diving gear, and the only sour note in the whole proceedings was certainly not his fault. He had been told to close a certain valve, which was in fact already closed, so, following instructions, he had actually opened it. Nevertheless his efforts had succeeded and soon the water level began to fall.

With the Great Spring conquered, work began to move ahead. Inevitably, new problems arose, including a strike where the workers demanded a reduction of shifts from ten to eight hours, but the contractors claimed they only worked eight hours anyway. At this late date, blasting was by dynamite, so after an explosion the men went back down the workings until the air was cleared and enjoyed a leisurely meal break. A new system was inaugurated: work from 6 a.m. to 9 a.m., when the first explosion was set off, then breakfast from 9 to 10 and so on through the day – work from 10 till 1, blast, dinner; work from 2 till 6, blast and leave. The works now were very different from the early days – compressed air drills were used and underground haulage was by ponies or small steam locomotives. As a result, accidents were few, but the men contracted rheumatism and pulmonary illnesses from the cold and the damp.

Work went on steadily, though there was a panic in 1882 when a leak from one of the subsidiary workings led to a rumour that the river had broken through, and men fled the workings. The following year, the real thing occurred. In vile weather, at Spring tides, a tidal wave drove up the Severn, swept on by gale-force winds. It overflowed the workings, dowsing the boilers of the steam pump, and swept down the shaft: three men who were climbing ladders to the surface were carried down to their deaths. As the water poured into the tunnel, it trapped eighty-three men who began a retreat in the face of the rising waters. They clambered up on to a wooden staging and waited for rescue. A boat lowered down the shaft appeared out of the gloom to carry them to safety. It was the last adventure – or rather misadventure – and in 1886 the tunnel was opened.

If the Severn tunnel was the greatest exploit of the Railway Age, at least it was not the most costly in terms of human life. None of the major tunnels were completed without this deadly price being paid. Sometimes, it was a single accident that caused the devastation, like that at the Watford tunnel on the London and Birmingham in 1835.[14] The night before the accident, the sub-contractor had been down the shaft to see the works. Quite near the bottom, the wooden supports were being removed to allow the bricklayers to get on with the permanent linings. It seems that the supports were moved too soon, for without any warning, the tunnel began to collapse around the foot of the shaft and the shaft itself started to cave in.

> The man who attended the Gin heard a cry of 'ware' that the ground fell in at the same instant and so suddenly that his Dog was buried in the abyss, the gin and Gear carried down & that he only escaped by being tangled in a part of the machinery above ground. That the candles of the men at work on the length tunnel in the next adjoining shaft on the north were blown out by the rush of air through the heading or driftway when the ground fell.

Ten men were named as dying in the fall – an eleventh, a labourer, went to his death anonymous and unnamed.

Major accidents made news, but throughout the railway workings there were accidents leading to injury, maiming and death which, if they were reported at all, were of purely local interest. The Bristol paper showed a keen interest in what was happening at the western end of the GWR: looking over the records for 1838 and 1839 a number of accounts appear of which these are a sample.

> On the night of Friday another fatal accident occurred at the shaft No. 5 of the Box Tunnel, to a poor man whose name is at present unknown. The deceased had just come up to the brink of the shaft, and was in the act of landing to take some refreshment, when his foot slipped, and he fell a depth of 260 feet. His body was taken to the Tunnel Inn, on the works, and presented a shocking appearance.

> On Monday, a man employed on the Great Western Railway works one Charles Harkison, went, with one of his fellow workmen to Biddestone and it is supposed had taken too much beer and laid himself down under a cart laden with soil, when it is conjectured some person or persons (of whom there is suspicion) wantonly upset the cart with its contents upon the unfortunate man and so crushed him that he died on Saturday. The deceased who was not 20 years of age was not addicted to drinking but was considered steady by all who knew him.

> A distressing event occurred on Friday at the open cutting in Keynsham Park, on the line of the Great Western Railway, by the falling in of a bank. One of the workmen was killed on the spot, and others were seriously injured. Three were taken to the Bristol Infirmary, viz., Isaac Gibbs, John Smith and Henry Hancock; the latter had sustained a very severe compound fracture of the left thigh, which it was found necessary immediately to amputate; the other leg was also fractured above the ankle. This man's recovery is very doubtful; the others had received slighter injuries, and will probably be discharged in a short time.

> We regret to state that one of those appalling occurrences which are almost unavoidable in public works of great momentum took place on Thursday afternoon at Brislington near this city in Tunnel No. 1 of the Great Western Railway. It seems that while some of the workmen were engaged in what is termed striking the centre of the arch; the brickwork gave way and overwhelming three of them killed them on the spot. A body of men rushed to their assistance when unfortunately a further mass gave way by which two of them sustained severe fractures of the skull and seven others were injured. An inquest was held yesterday on the bodies of the deceased and a verdict of *accidental death* returned. The two wounded men who suffered most were conveyed to the Infirmary in a dangerous state.[15]

In 1841, the Bristol Infirmary reported that the number of casualties was 'unusually high', but gave no figures. Perhaps the doctors shared the view of Mr Henry Lacey Pomfret of Hollingworth during the construction of Woodhead tunnel, who recorded 23 compound fractures, 74 simple fractures and 140 miscellaneous injuries including burning and blinding, but when asked about deaths replied that he had no records, since the dead were not requiring the services of a surgeon.[16] The same doctor declared that he had personally never heard of any enquiries being made after accidents which did not end in death – amputations and the like did not require any special measures. It has to be said that official reports do on occasion lack credibility. One young man, coming up from a night shift in Bramhope tunnel, took the wrong turning in the dark and instead of finding the path home only discovered the open mouth of the next shaft, down which he plunged to his death: 'No one saw him fall in but heard a cry of O-dear.'[17] Whatever one might imagine a burly young navvy saying as he toppled hundreds of feet down a tunnel shaft, one cannot somehow believe in the official 'O-dear'.

In those days, long before any National Health Service existed, someone had to pay for the treatment of the maimed and injured. Statistics are scarce, but the Northampton Infirmary recorded that it treated a total of 124 cases of navvies working on the London and Birmingham between 1835 and 1839.[18] Similar numbers were reported from other lines.[19] But the Northampton hospital also gave an estimate of the costs and who paid them: £115 10s. paid by the company, £10 10s. by the contractors and £4 11s. by the workers themselves – yet the estimated costs came to £597. Who paid the outstanding £466 9s.? Nothing is said, but the inference is that the hospital carried the costs itself – certainly there are no records of any injured navvy being turned away for want of funds. The navvies did have their own sick funds, but how much was paid in and how far, in practice, it met hospital expenses is uncertain. In any case, the sick-clubs were not always supported. Captain Moorsom was one engineer who tried to improve safety at the works, particularly in the use of gunpowder, and was able to record that in all but two of the contracts in the works sick-clubs had been set up, but in these two cases no one could be persuaded to act.[20] Peto told the 1846 Committee that he added to the navvy's sick-club funds without telling them. When a man was ill, payments of 8 to 12 shillings a week were made, but in the case of an accident Peto paid compensation himself. Few contractors matched his generosity.

It is easy when writing about railway construction to think of the work of the navvies and forget the many tradesmen who worked alongside them. In tunnelling we have already come across the miners who arrived from the coalfields or the metal mines of the south-west, but sometimes a few words in the official documents bring home the reality. The simple statement 'the tunnel was lined' gives little indication of what was actually involved. Here is Robert Stephenson giving proposals for carrying on work at Kilsby tunnel:

> To erect a steam clay mill with kilns etc sufficient to supply 30,000 bricks per diem: say total quantity of bricks required 20,000,000 then 10,000,000 to be made the first season, and supposing 6,000,000 may be obtained from the open cuttings at the two ends of the tunnel and the neighbouring brickworks, 4,000,000 will be required from tunnel clay, then average number of working days in the season, say 15 × 30,000 = 4,500,000.[21]

Now this represents quite a large industrial enterprise – setting up and running a small brickworks, carrying in the raw material and taking out the finished bricks, and employing enough bricklayers to lay them in the dark, difficult surroundings of the tunnel at the rate of 30,000 per day. Tunnelling suddenly seems to have become something more than a lot of men making a hole in the ground.

In other specialist areas, other tasks were going forward. Saw-pits were established to provide timber for the railways. In these the log is laid over the pit and sawn up using a massive two-man saw. One man stands above the log and pulls the saw up, his partner – inside the pit, permanently showered in sawdust – pulls the saw back down. Carpenters then worked on these sawn-up logs to provide the wooden centerings, for example, on which the arched roof of the tunnel was constructed. Centerings were also used for the vast numbers of bridges which carried the tracks over or under roadways, provided access for a farmer to his fields, and crossed streams, rivers and canals. Built of brick or stone, iron or even, more rarely, wood, they scarcely attract any attention, but they were as important as any other structures along the line and contractors would give quite precise specifications of the standards to be met. It was an area where economies could be made. A company might have made extravagant promises to landowners about what they would supply as the railway passed through their land, but once work began things could look very different: 'Private roads which were said to require archways of great height to allow Hay Waggons to pass are bought up for a trifle and the embankment which was to pass over them may be lowered.'[22] Even so, there was enough work to keep an army of carpenters, bricklayers and masons busy.

The numerous small bridges that dot every line in the country represented a great effort, and are often surprisingly interesting. There are fine examples of bridges built on the skew to accommodate a road that crossed the tracks at an angle – the most famous being the skew bridge at Rainhill, near where the *Rocket* steamed on its triumphal runs at the trials of 1829. Other railway companies made life easier for themselves by building the bridges square on and realigning the road to zig-zag across through two right-angled bends – a device which now irritates motorists who travel anywhere near the route of the Cambrian Railway in Wales. But the structures that capture the imagination are the great bridges and viaducts.

Just as railway companies used their more important stations to make statements about the grandeur of the railway as a whole, so too the great bridges and viaducts were more than just parts of a transport route. Balcombe viaduct which carries the London to Brighton line over the Ouse valley is one of the best examples. Above its thirty-seven tall arches, the track is guarded by a neat balustrade, while stone pavilions add a touch of grace at both ends. For the scale and monumentality, praise is due to the engineer John Rastrick; the elegant touches were the work of architect David Mocatta. Very rarely has a record been left of most of the men who worked on such constructions. However, the builders of the viaduct at Yarm were so pleased with their work that they left an inscription on a pier high above the River Tees, hardly the easiest place to read it. It gives the names of the engineers Thomas Grainger and John Bourne, the Superintendent Joseph Dixon, and the contractors Trowsdale, Jackson and Garbutt. It also provides the information that a total of 7 million bricks were used. The work was completed in 1849, ending a task that had begun in 1825 when the Stockton and Darlington Railway arrived at the north bank of the river, and then was delayed for a quarter of a century before being allowed on its way to Northallerton.

The system for building the great viaducts of brick and stone was no different from that used on road bridges – only the scale altered. At Yarm, the road could come down the hill, hop across the river on a low bridge, and climb up the opposite bank. The railway, needing to keep on the level, had to stride over the rooftops of the town before reaching the river – fifty-three arches were built in total. Yet the road-builders perfectly understood what the railway builders were doing. Piers were built up – in the river this entailed creating a dry area behind coffer dams, but otherwise it was simply a question of building a series of brick towers. The piers were then joined by centerings, wooden arches over which the bricks were laid. Then, when the mortar had set, the centerings were removed and the arch stood firm – in theory. Some proved firmer than others.

The Rother viaduct on the Manchester, Sheffield and Lincolnshire Railway was under construction, with twenty of the thirty-six arches completed, when in October 1848 rain brought floods that swirled around the site. Orders were given to shore up the nineteenth arch, but even while the men were at work, the arch collapsed, killing four men. Then, like dominoes, the rest followed: 'In a few minutes a dozen more of the arches followed the nineteenth in regular succession, the noise and concussion resembling those accompanying an ordnance or an earthquake.'[23] The disaster could be attributed to misfortune: in other cases it could be put down to poor workmanship or to the disease that plagued so much work – the need to rush on with the work to meet deadlines or get payments. A disaster that occurred very late on in railway building history took place on the Stanway viaduct between Cheltenham and Honeybourne in 1902. The contractors were removing centering from the arches when one arch gave way, to be followed by three others. Two factors caused the disaster, in which four men died: first, and most importantly, the mortar was not set. This might not have been a problem, but to help remove the woodwork a heavy steam crane had been trundled on to the viaduct adding to the strain on the weakened arch. The report later criticized the mortar used as well. It was, in short, an accident that need never, and should never, have happened.[24]

One other traditional material found favour with engineers – timber. It was widely used by Brunel on the lines that ran down through South Devon and Cornwall, and for once no one could claim that this was one of his strange, quirky notions. Timber was used for the very sound reason that it was cheap, and the lines could never expect the sort of revenue that would appear on the main line between Exeter and London. By the end, Brunel had worked out a system of standardized bridges with just two spans – 50 ft. for short, 66 ft. for long – that would meet most needs. They were cheap to build and easy to maintain. These beautiful, yet fragile-seeming structures were carried on stone piers. Today, sadly, only the piers remain, like rotting stumps, while the trains run over new and less exciting bridges. Of all the great timber bridges, the most splendid to survive is the 800-ft.-long structure at Barmouth, with an opening span in the centre to let ships through.

The eighteenth century had seen the first iron bridge being built, and the railways took to the new material with some enthusiasm: Stephenson at Menai, Conwy and the Tyne, Brunel at the Tamar and many others. Towards the end of the century a new material was brought in for what is perhaps the most famous railway bridge in Britain, the Forth Bridge.[25] The men in charge were sound practical men and experienced engineers – Sir John Fowler and Benjamin Baker (who was also to be knighted for his work on the bridge). The contractor on the bridge, William Arrol, was also to be knighted. Ironically, this bridge, often spoken of as one of the triumphs of British engineering, relied during the early years of construction on foreign workers. The key to the whole structure lay in the little island of Inchgarvie which formed the base for the central pier, from which the cantilevers reached out towards the shore. Incidentally, Baker never referred to this as a cantilever bridge at all but as a 'continuous girder bridge'. The islet however was not large enough to take all the supports, so a good part had to sit on a submerged ledge. To do this caissons were used – giant metal cylinders 70 ft. across and as much as 90 ft. high, depending on the depth of water in which they were to be used. These great cylinders had a lower cutting edge that bit into the seabed to hold them in position, and a compartment at the bottom kept free of water by compressed air. Here the men worked preparing the foundations. As it was a technique developed and widely used in Europe, the work was entrusted to a French company who brought over Italian workmen. At Queensferry the caissons had to be sunk as much as 89 ft. below high-water mark, dropping through silt and soft mud before reaching a solid base. Work in the caissons must have been unpleasantly claustrophobic, and the compressed-air atmosphere produced what became known as 'caisson disease', a condition not unlike the diver's 'bends'.

The men, the 'briggers', who built the great towers were mainly Scots with experience in the engineering industry – men with knowledge of working with metals, and who understood the basic construction technique of riveting the plates together – even if they had less experience of working a couple of hundred feet up in the air. Some 4,000 men came to work on the giant bridge and 57 died before it was completed. Inevitably the blame was put firmly on the men, who were reckless. Like many men in dangerous jobs they were proud of their ability to wander about on the high scaffolding and girders with the casual ease with which the rest of humanity walks down a city street. Drink was a problem, and so too was carelessness – even a single rivet dropped from that height could kill. The human price paid for the bridge was as high as in any of the great tunnels which attracted so much attention earlier in the century. Drunk, reckless or careless they may have been, but the briggers produced a bridge described in the official report as 'a wonderful example of thoroughly good workmanship'.

Grand structures attract attention – it is as much a thrill to cross the Forth Bridge now as it was a century ago – but for every line that could boast some immense, spectacular feature there were a dozen of more modest appeal. And even the grand lines had all the small details that were as important to success as the more flamboyant. What might seem an irritating nuisance to the railway company could be a matter of great concern to local people such as the group who petitioned for little tunnels to be left under the embankment at Coatham Marsh so that rabbits could get through. This was not an early example of ecological conscience, but rather the result of the desire of the farmers on one side of the bank to have as many rabbit pies as those on the other. The fact that the same company proposed building a gas-works that would not only light the new station but would also supply enough gas to light the whole town as well, was far less important as far as they were concerned.[26]

With the ground prepared, the last phase of construction got under way, setting out of the permanent way – laying sleepers, spiking in chairs and rails, and ballasting the track to hold everything firm. There was signalling to set in place, wayside halts to build, goods sheds and engine sheds to construct, water towers to put up for the thirsty engines; and houses to be found for locomotives and rolling stock. The sheer multiplicity of tasks that had to be done between the day when the first sod was cut and the day when the first train made its triumphant progress is difficult to comprehend. It required organizational skills, specialist knowledge and experience, and the work of hundreds and often thousands of men. No wonder that, however modest the line, the opening day was a cause for celebration.

CHAPTER TWELVE

The End of an Era

You may travel by steam, or so the folk say,
All the world over upon the railway.

Song written for the opening of the Grand Junction Railway

Within half a century of the opening of the Stockton and Darlington Railway, the railway map of Britain was largely filled in, with the exception of a few areas such as the west of Scotland. Thousands of miles of track spread throughout Britain, and in that half-century almost nothing had changed as far as the art of construction was concerned: mechanization was scarcely any further advanced in 1875 than it had been in 1825. The one really new device that did come into use during the period was the steam hammer, invented by James Nasmyth. It had a curious early history, involving one of the great railway engineers, but not the railways. When work began on Brunel's famous iron ship, the SS *Great Britain*, it was intended, like all other big steamships of the age, to be driven by paddles. The shaft was to be a massive 30 in. in diameter, and Nasmyth devised what was by far the most powerful hammer in the world to forge it. Then Brunel was converted to the idea of the screw propeller, so the paddle shaft was never needed. Robert Stephenson, however, found that the new steam hammer was ideal for pile-driving for his Royal Border Bridge, across the Tweed, which was opened in 1849.

As the century drew to a close, the gaps in the railway map were gradually filled in – one such gap was closed by what is perhaps the most attractive of all British lines, the West Highland branch from Fort William to Mallaig. Here, a new material came into use, concrete. It can be seen in bridges and viaducts and at its most spectacular in the Glenfinnan viaduct. Some 127 ft. long and carried on twenty-one arches, it swoops round in a tight curve which, if it does nothing else, delights modern enthusiasts who can poke their heads out of the carriage windows and take a photograph of their own train crossing the bridge. The material was new, but the construction itself was wholly traditional. Robert McAlpine simply treated blocks of concrete as if they were blocks of stone and piled them up in arches just as his predecessors had done. The same could be said of another equally spectacular concrete viaduct across the Tamar at Calstock. It is certainly a magnificent structure, of even greater proportions – 1,000 ft. long and rising over 100 ft. above the river. But again techniques looked back not forward. The contractor J.C. Lang of Liskeard set up works on the Devon bank, and just as nineteenth-century contractors opened quarries and shaped stone blocks, so he set about turning out concrete blocks – 11,148 of them. The viaduct did have one novel feature – in order to join the railway to the quay far below, a vertical wagon lift was added to the side of the viaduct. It was opened in 1907 – the year the first pre-stressed concrete bridge was constructed, not in Britain, but in France.

The Calstock viaduct has another claim to fame, however. The engineer responsible was Colonel Holman Fred Stephens, the man associated more than any other with the last great flowering of railway building in Britain.[1] The Colonel's background was not that which one normally expects to find among engineers – his father Frederick Stephens was one of the foremost art critics of the day, a great friend of the Pre-Raphaelites, hence the name Holman, after Holman-Hunt. If the young Stephens had an artistic soul he kept it well hidden; the extravagance and exuberance that ruled the Pre-Raphaelite set found no echo in the young man's heart. His goal was not purity and beauty, but economy. He was the right man growing up at the right time. He was born in 1868, studied civil engineering at the University of London and mechanical engineering with the Metropolitan Railway. He began his working career as resident engineer on the little Paddock Wood and Cranbrook Railway, a line of the type that was to be regarded by Stephens as the ideal for a little country line. It had sharp corners, steep gradients and stations cobbled together out of old timber and corrugated iron, but it was cheap to build and cheap to run. And this was what was needed to fill in the gaps on the railway map. Cost was the great obstacle to progress – the cost of obtaining an Act of Parliament, the cost of building a line to meet regulations designed for the safety of a fast, main-line express service. One solution was to build to different specifications. Experiments on the Festiniog Railway in the 1860s had shown that narrow-gauge lines – in this case a mere 2 ft. – built for the horse-drawn tramway age with severe gradients and corners could be worked by steam locomotives. Such lines were far cheaper to build than standard-gauge ones. Parliament then eased the way for standard gauge gap fillers with the passing of the Light Railways Act of 1896.[2] Now, instead of an Act, a line could be built under an Order granted by a three-man commission. Standards were lowered, but a price had to be paid: the maximum speed was set at 25 m.p.h. Still, many communities felt that 25 m.p.h on a train was better than 3 m.p.h on foot, and queues of would-be railway builders were soon forming. It was now that Stephens came into his own, running a whole network of lines up and down the country from a modest office at a suitably modest address – 123 Salford Terrace, Tonbridge.

The light railways often seemed to mimic the history of their greater brethren in miniature. The Lynton and Barnstaple Railway is a classic example of enthusiasm triumphing over common sense in a way that had not been seen on the railways since the heady days of the mania.[3] The line was to run from the Devon and Somerset Railway at Barnstaple and skirt round the edge of Exmoor to arrive at Lynton. It was unfortunate that the customers wanted to arrive at Lynmouth at the edge of the sea, not at Lynton, which stood high on a hill above it, but there was nothing to be done. The line was duly authorized in 1895, and an optimistic prospectus was issued. It proclaimed that thirty-four coaches a day arrived at Lynton, but it failed to mention that they came from nearby Ilfracombe, not distant Barnstaple; the station 'has been chosen so that it will not be visible from Lynton or Lynmouth', and, it might have added, not very accessible either. But of all the statements put out at the time none were wider of the mark than this: 'The cost of construction would not exceed the sum of £50,000, the line contained no works of any great magnitude, being almost altogether a surface line, would be open for traffic in the summer of 1897, and be completed and equipped without entrenching on the Company's borrowing power.' Not one of those confident predictions was to be fulfilled.

There was trouble right from the start. The company was confident that it had the support of local landowners. So it did, but that did not prevent those same supporters from demanding, and getting, what the company described as 'extortionate' prices for their land. Inroads into the company's funds were soon being made, and work had yet to begin. In looking at the estimates for the line, it is difficult to decide who was the more foolish, the engineer Sir James Szlumper who blithely reported that it was simply a matter of cutting a line through surface soil, or the contractor, J. Nuttall of Manchester, who accepted his opinion and took a fixed-price contract with no let-out clauses whatsoever. A glance at the rocky terrain might have suggested that the going would not be so easy as the optimistic report supposed, and digging out a flat surface on the side of a steep hill would necessarily be a troublesome business. The result was inevitable: costs escalated and Nuttall asked for a further £27,000. Sir James Szlumper, exchanging his contracting engineer's hat for that of arbitrator, promptly agreed. The company, however, objected and the courts decided in their favour. The company won, but it was a hollow victory, for all they got was a bankrupt contractor and an unfinished line. The familiar story was being played out yet again in the new environment of the light railways. And this story had no happy ending. The line opened in 1898, only to be closed in 1935.

The light railways had a quirky charm all their own and their idiosyncrasies earned them many romantic admirers. Some served outlying rural communities well; some, founded on nothing but wholly unjustifiable opinion, foundered almost before they had begun. The resoundingly named Bideford, Westward Ho and

Appledore Light Railway began with high hopes in 1901 and sold the last ticket in 1917. Some lasted long enough to be taken up, or revived, in the modern preservation movement. But their construction left few epic tales to be told – indeed, they attracted very little attention, even locally. But then this was true to a certain extent of the main-line railways of the earlier age. Papers reported accidents; riots made news; and in the 1840s the excitement of share dealing resulted in columns of railway statistics appearing as a regular feature. Yet surprisingly few people seemed to want to go and see, or describe, the great works that were going on throughout the land. Even when visitors did go to see the works in progress, their accounts are often somewhat less than informative. In 1836, the British Association gathered in force at the invitation of the Directors of the Great Western to view the works. As the line was still being built they took to the water and the River Avon. A procession of boats, some sixty in all, set off on a damp Friday morning from Bristol. Flags were flying, a band was playing and crowds lined the bank as the procession set off with Commodore Claxton in the lead.

The party landed near Bath and proceeded to walk down two short tunnels and along a length of track. The view from the hill above the tunnel was remarked upon: 'The Philosophers appeared to be highly pleased with the view.' And that, as a description of one of the greatest engineering enterprises of the age, was that. An explanation as to why so few details of the trip were recalled might be found in the account of the boat journey back from Hanham.

> A friend near us said 'I'm sure the Commodore won't be at fault; he knows 5 miles march will make us thirsty, if not hungry.' Our friends were right; for upon looking into 3 boats which Captain Claxton had in the procession we found an immense number of parcels and the bottom of the boat well laden with bottles of wine, porter and cider. – bravo! Commodore, said we. The boats were then ordered to be moored and each boat as it rowed alongside the commodore had several of these parcels handed in which upon examination were found to contain sandwiches and biscuits. These were very soon devoured; for although Philosophers delight very much in theory, a little substance occasionally

> is not amiss. We then by order bent our boats nose homewards. The Commodore again took the lead and when he thought all were well under weigh he gradually dropt astern; but as each boat came up the silver cups were handed out in so tempting a manner that for the life of us we could not avoid stopping to look at them – and having so stopped we found we must drink health and long life to Commodore Claxton; this was well done.[4]

The lack of interest in the construction of the railways was partly due to the fact that the novelty had begun to wear off. There was little remarkable in finding a railway gang at work in the countryside in the 1840s – to some it seemed remarkable only to find an area that was uninvaded. It was also the case that many of the more dramatic scenes took place in remote and inaccessible areas. In the early days it was not the men digging the line that created the excitement, but the amazing new machines that would use it. The young actress Fanny Kemble, then at the height of her career but soon to depart for a much less happy life on an American slave plantation, was treated to a footplate ride along the Liverpool and Manchester Railway with no less a character than George Stephenson at the regulator. She described in vivid terms the effect of passing through the deep Olive Mount cutting:

> You can't imagine how strange it was to be journeying on thus, without any visible cause of progress other than the magical machine, with its flying white breath and rhythmical, unvarying pace, between these rocky walls, which are already clothed with moss and ferns and grasses; and when I reflect that these great masses of stone had been cut asunder to allow our passage thus far below the surface of the earth, I felt as if no fairy tale was ever half so wonderful as what I saw. Bridges were thrown from side to side across the top of these cliffs, and the people looking down upon us from them seemed like pygmies standing in the sky.

But her greatest admiration was reserved for the wonderful new engine – and the man who drove it, who had 'certainly turned my head'.

> The engine having received its supply of water . . . was set off at its utmost speed, 35 miles an hour, swifter than a bird flies (for they tried the experiment with a snipe). You cannot conceive what the sensation of cutting the air was; the motion is as smooth as possible, too. I could either have read or written; and as it was I stood up, and with my bonnet off 'drank the air before me'. The wind, which was strong, or perhaps the force of our own thrusting against it absolutely weighed my eyelids down. When I closed my eyes this sensation of flying was quite delightful, and strange beyond description; yet strange as it was, I had a perfect sense of security, and not the slightest fear.[5]

Others shared her enthusiasm, and it is probably as true today that for every one who admires some great feat of civil engineering, there are a hundred who thrill to the hiss of steam, and the clatter of wheels over rails. The crowds that turned out for the grand opening ceremony saw the locomotive as the star attraction. This was certainly true on the first grand opening, the Stockton and Darlington's great day of 27 September 1825.[6] It was all carefully planned, and an official announcement was made: 'A superior locomotive, of the most superior construction, will be employed with a train of convenient carriages, for the conveyance of the proprietors and strangers. Any gentleman who may intend to be present on the above occasion will oblige the Company by addressing a note to their office in Darlington as early as possible.' The 'superior locomotive' was *Locomotion*, with George Stephenson and his brother James on the footplate and Timothy Hackworth as guard. The convenient carriages were, in fact, converted coal trucks, though space was found for the company's one coach 'The Experiment'. Three hundred tickets were issued and everything duly ordered – proprietors, engineers, surveyors, workmen and the merely curious, all were allotted their places. In the event it was all a shambles. The crowds milled round the trucks, those who had tickets tried to claim their seats and those who had not tried to get in anyway – and if they could not get in, then they hung on to the outside. The bands played, church bells rang and crowds gathered all along the route, as the train made a steady if slow progress along the new line from Shildon to Stockton-on-Tees. The day ended with a huge feast and a great many toasts, including one to the other sturdy infant railway, the Liverpool and Manchester. It was the thirteenth toast of the night, and the superstitious for once could claim that they were justified in prophesying gloom and doom.

The Liverpool and Manchester was, or should have been, the grandest opening of them all, the first inter-city route, the line which had seen the trials for the locomotives. There were to be seven of them in steam that day in September 1830 and the guest list included foreign princes and the famous Duke of Wellington, hero of Waterloo. Unhappily, in the north of England, he was more generally regarded as the infamous Duke of Wellington, villain of Peterloo. He was an odd choice as principal guest, since he made no secret of his lack of enthusiasm for railways, which he declared only 'encourage the lower classes to travel'. Furthermore, he had argued openly and publicly with the local MP and enthusiastic supporter of the line, William Huskisson. Everything was to be done in the grandest style, and that included ordering ornate carriages, like Baroque summer-houses on wheels, for the principal guests. If the organization of the Stockton opening was shambolic, then that for the Liverpool and Manchester was even more so. Arrangements had been made for the ducal party to stop at Parkside where a grandstand had been erected so that they would see the grand procession of locomotives go by on the other track. But, instead of taking their seats, the dignitaries wandered around as if it were Hyde Park, not Parkside. Huskisson had gone across to accept a reconciliatory handshake from the Duke when there were shouts that a train was coming. Most of the party scrambled into the carriages, but Huskisson alone seemed bewildered: the advancing *Rocket* hit the carriage door, hurling him on to the track where a wheel ran over his thigh. In spite of a high-speed dash of mercy by *Northumbrian* to the nearest medical help at Eccles, the railway's great supporter died on the opening day.

The two events show, if nothing else, that railways were still a novelty. Spectators crowded to see the new steam locomotives and few, as Huskisson's death so tragically showed, had the remotest idea of what the new age meant in terms of speed and power. As the railways spread, the novelty factor diminished, but not the enthusiasm of local people, joined to the railway network for the first time.

Travel was brought within the means of many who could never have afforded it before. Parliament even legislated for the poorer travellers – a rare departure from the *laissez-faire* economics of the age – in a bill of 1844, which contained a clause 'to secure to the poorer classes of traveller, the means of travelling by Railway at moderate fares, and in carriages in which they may be protected from the weather'. These 'Parliamentary trains' had to cover the whole route at least once a day, and they were to charge the third-class passengers no more than a penny a mile. The railway companies showed their enthusiasm by running the trains with their slowest engines and oldest rolling stock, but for people who had never been able to afford to travel any distance beyond that to which their two feet could take them, the railways opened up the world. So they came in their thousands, and the first arrival at any town was an excuse for a grand ceremony.

> The procession was formed about two o'clock headed by Mr Wilday followed at a respectful distance by the 'navies' each supplied with a barrow and spade and also with a new hat . . . The different orders of Oddfellows, Druids and Freemasons followed next with their respective bands. A large number of the Charity School children brought up the rear. The scene of action was a field belonging to Wilday in the middle of which he stood while the procession formed around him. The youngest children were in the first circle, the next tallest in the 2nd and so on then the different clubs. At a given signal all the navies wheeled their barrow-fulls into the circle and emptied them so as to form one large mound amid the cheers of the populace.[7]

Meanwhile, seventeen sheep were being roasted at the local pubs and, as the account concludes, 'All went to bed fully satisfied.'

Railway guide-books appeared to help the new generation of travellers. One guide published in 1851 had a curious format.[8] On one page it described the view out of one window; on the facing page, there were details of the panorama opening out on the opposite side – to keep life simple, it was as well to sit facing the engine on the outward journey and with one's back to the engine on the return, so the left-hand page and left-hand view coincided. It gave immense detail, right down to the names of the owners of the grand houses seen along the way. What it did not do was say anything along the way about the railway itself – that was limited to an introduction. One line, however, never lost an opportunity to blow its own trumpet. The Great Western was also the Great Publicist, and remained so throughout its existence. An *Illustrated Guide* simply announced that passengers would want to be informed of all the details of the magnificent line along which they were speeding.[9] As they rattled into Box tunnel, they were showered with facts and statistics.

> Its entrance from the east is through a deep cutting with vertical sides hewn out of the rock; the arch is semi-circular, formed of rusticated work, and springing from upright sides; and on either side is a projecting pier. As respects the rocks through which the tunnel penetrates, we have already remarked . . . that it traverses in succession the great or middle (sometimes called the *upper*) oolite, so well known as the Bath building-stone, – then two thick beds of fuller's-earth and light clay, – after that the lower or lesser oolite, – lastly, a blue marl-stone or shale, separating the oolites from the lias beneath; and it is in this stratum that the tunnel terminates, and pursues its course all the way to Bath. The Box-tunnel is 3,195 yards, or about a mile and three-quarters long, lined nearly throughout with brickwork, and ventilated by six shafts, each 25 feet in diameter, and varying from 70 to 300 feet in depth; the whole run through it being a rapid incline westward. This tunnel, it may be interesting to know, occupied two years and a half in formation, involving the labour of excavating no less than 414,000 cubic yards of earth and stone, chiefly the latter; besides the construction of 54,000 cubic yards of masonry and brickwork, with a consumption of more than thirty millions of bricks. A ton of gunpowder was used weekly in blasting, and a ton of candles for lighting the labourers, who averaged more than a thousand during the whole undertaking.

Quite how the travellers speeding through the dark tunnel were expected to pick out details of oolite and shale and fuller's earth was not made clear. But in the flurry of statistics, there was just one brief mention of the men who made it all possible – no mention of Brunel, nor of his assistants, and certainly no names were given to any of the great army who worked and, in sadly too many cases, died in creating the tunnel. It has been estimated that someone died for every mile of track that was laid as the Great Way West spread its tracks from London to Bristol. If there are ghosts to be seen as the carriage lights flicker over the tunnel walls, then they are the ghosts of brawny young men, hats at jaunty angles, clay pipe in mouth and shovel or pickaxe slung nonchalantly across a shoulder. Perhaps it was enough that the work itself was there to be admired.

The opening day was a grand event for the local people who came to watch, or if they were lucky, ride the new trains, in their smart, freshly painted liveries. It was closing day for many more. They were out of work. Engineers looked for other lines to build; assistants assiduously applied for posts higher up the professional ladder but were often glad enough to take whatever came their way. Contractors prepared fresh bids for new works. The navvies simply moved on, their worldly goods strapped to their backs, to find fresh work. Some, employed locally from the first, went back to the land or the mines or wherever they had made their living before the railway came with its promise of high wages. Others went on the tramp, stopping overnight under a hedge or, if they were lucky, in the local workhouse. Word was passed on about diggings across the country and if no work was to be had there was a tradition among the navvies who did have work of helping mates on their way. On they walked until they found a spot where strong muscles and hard-won expertise were required. There they began all over again – digging another railway.

Chronology

This is not intended as a complete chronology of the Railway Age, but rather as a list of key dates in the nineteenth century.

1801 The Surrey Iron Railway is incorporated, the first railway line to be sanctioned by Parliament. It used horses for haulage.
1804 Trevithick successfully demonstrates his steam locomotive on the Penydarren Tramway.
1812 The opening of the Middleton Colliery Railway.
1814 George Stephenson's first steam locomotive, *Blücher*, is set to work.
1825 The opening of the Stockton and Darlington Railway.
1829 The Rainhill Trials test steam locomotives. The *Rocket* wins.
1830 The opening of the Canterbury and Whitstable Railway and the Liverpool and Manchester Railway.
1831 The Garnkirk and Coatbridge Railway opens – the first steam railway in Scotland.
1837 The opening of the Grand Junction Railway and the London and Birmingham Railway.
1838 The London and Greenwich, London's first public railway, opens throughout its length.
1840 The first trials of the atmospheric railway in Britain take place.
1841 The Taff Vale Railway is opened. The Great Western Railway opens from London to Bristol.
1844 The Midland Railway is created by amalgamation of the North Midland, the Midland Counties and the Birmingham and Derby Junction Railways.
1845 The Woodhead tunnel on the Sheffield, Ashton and Manchester Railway is completed.
1846 The height of the Railway Mania: 272 Acts of Parliament are

passed. The London and North Western Railway is formed.

1847 The Caledonian Railway is opened.

1848 The Chester and Holyhead Railway, including the Menai Straits and Conwy bridges, is opened.

1850 The Royal Border Bridge across Tweedmouth is opened.

1854 The North Eastern Railway is formed by amalgamation.

1857 The first steel rails are introduced.

1859 The Royal Albert Bridge across the Tamar is opened.

1863 The Metropolitan Railway, the first underground railway, is opened.

1876 The Settle and Carlisle Railway is opened.

1878 The Tay bridge is opened in May: it collapses in December.

1886 The Severn Tunnel is opened.

1890 The Forth Bridge is opened.

1892 All broad-gauge track is converted to standard gauge.

1896 The Light Railways Act is passed.

1897 The Great Central Railway is opened.

Notes

JRCHS = *Journal of the Railways and Canals Historical Society*

CHAPTER 1

1. Darlington and Barnard Castle Railway Papers, PRO.
2. Letter, Richard Trevithick to Davies Giddy, 4 February 1804.
3. Quoted in J. Bushell, *The World's Oldest Railway*, 1975.
4. Letter, Robert Batcherby to Richard Miles, 20 August 1818.

CHAPTER 2

1. Quoted in Richard S. Lambert, *The Railway King*, 1934.
2. EMSP, *The Two James and the Two Stephensons*, 1861 (reprinted 1961). The author was almost certainly William James' daughter, Mrs Paine.
3. EMSP, *The Two James and the Two Stephensons*.
4. John Francis, *A History of the English Railway*, 1851.
5. EMSP, *The Two James and the Two Stephensons*, Letter from Joseph Sanders, 25 May 1824.
6. John Francis, *A History of the English Railway*.
7. EMSP, *The Two James and the Two Stephensons*.
8. Committee Enquiry into the London and Birmingham Railway Bill, 2 May 1839.
9. J.B. Latimer in *The Annals of Bristol*, 1887.
10. Sheffield and Goole Railway Prospectus, 13 September 1830.
11. Report of the public meeting held on 24 November 1835 at Colchester.
12. John Francis, *A History of the English Railway*.
13. *The Annals of Bristol*.
14. A full account of Hudson's career can be found in Richard S. Lambert, *The Railway King*, 1934.
15. D.M. Evans, *Facts, Failures and Frauds*, 1859.
16. Eastern Counties Railway, Miscellaneous Papers, PRO.

17. D.M. Evans, *Facts, Failures and Frauds*.
18. *The Railway Times*, 28 April 1849.
19. D.M. Evans, *Facts, Failures and Frauds*.
20. Frederick S. Williams, *Our Iron Roads*, 1852.
21. Olinthus J. Vignoles, *Life of Charles Blacker Vignoles*, 1889.
22. *Felix Farley's Bristol Journal*, 22 February 1845.
23. M.C. Read, *Investment in Railways in Britain*, 1975.
24. *The Railway Times*, 26 May 1849.
25. *Railway Intelligence*, September 1853.
26. *The Railway Times*, 3 February 1849.
27. Letter from J. Whitewell, 11 November 1857.
28. *JRCHS*, Vol. XXVII, No. 6, 1982.
29. Wycombe Railway Papers, 1856, PRO.

CHAPTER 3

1. Evidence put to the Select Committee on Railway Bills, 1836.
2. EMSP, *The Two James and the Two Stephensons*, letter to John Cawood, 26 December 1824.
3. Peter Lecount, *A Practical Treatise on Railways*, 1839.
4. I.K. Brunel letters, 1834 (undated).
5. Letter, October 1824, quoted in Anthony Burton, *The Rainhill Story*, 1980, in a full account of the early years of the Liverpool and Manchester Railway.
6. Samuel Smiles, *The Lives of George and Robert Stephenson*, 1874 edition.
7. *Felix Farley's Bristol Journal*, 30 August 1845.
8. Darlington and Barnard Castle Railway Papers, PRO.
9. T. Mackay, *The Life of Sir John Fowler*, 1900.
10. A full account can be found in *JRCHS*, Vol. XIX, No. 1, 1973.
11. T. Baker, *Railway Engineering*, 1848.
12. T. Baker, *Railway Engineering*.
13. I.K. Brunel, Private Letter Book, 15 April 1836.
14. F.R. Conder, *Personal Recollections of English Engineers*, 1868 (reprinted 1983).
15. William Fairbank Papers, PRO. Letter dated 3 October 1844.
16. William Fairbank Papers, PRO. Letter dated 10 October 1844.
17. T. Mackay, *The Life of Sir John Fowler*.

18. A full account can be found in John Thomas, *The West Highland Railway*, 1965.
19. Grand Junction, Great Western and South Western Junction Railway Letters, 27 November 1845 and (undated) December 1845.
20. T. Mackay, *The Life of Sir John Fowler*.
21. Olinthus J. Vignoles, *Life of Charles Blacker Vignoles*.
22. Quoted in Gordon Biddle, *The Railway Surveyors*, 1990.
23. Frederick S. Williams, *Our Iron Roads*, 1852.
24. *Felix Farley's Bristol Journal*, 17 October 1845.
25. Frederick S. Williams, *Our Iron Roads*.
26. *Annual Register*, 1845.
27. *Bristol Railway Estimates Book*, 1833–4.

CHAPTER 4

1. Frank A. Sharman, 'Colonel Sibthorp', *JRCHS*, Vol. XXVII, March 1982.
2. Frederick S. Williams, *Our Iron Roads*.
3. *Annals of Bristol*, 1887.
4. Quoted in R.H.G. Thomas, *London's First Railway*, 1972.
5. Quoted in R.H.G. Thomas, *London's First Railway*.
6. For example, in a petition against the Leeds and Thirsk Railway extension to Pateley Bridge, PRO.
7. John Norris, 'Railways to Oxford', *JRCHS*, Vol. XXX, November 1991.
8. See Anthony Burton, *The Canal Builders*, 1972.
9. *Felix Farley's Bristol Journal*, 29 August 1846.
10. Great Western and Wycombe Railway Extension Papers, PRO.
11. John Francis, *A History of the English Railway*. Here, he is describing the Great Western Railway.
12. I.K. Brunel, Private Letter Book, 26 January 1836.
13. Parliamentary Papers, 1845, Vol. X.
14. John Ruskin, *Praeterita*, 1885–9.
15. Peter Lecount, *A Practical Treatise on Railways*, 1839.
16. South Eastern Railway petition, quoting a letter dated 3 February 1845.
17. Liverpool and Manchester Railway: Minutes of Evidence, March–May 1825.
18. London and Birmingham Railway Bill, 2 May 1839.

19. T. Mackay, *The Life of Sir John Fowler*.
20. Select Committee on the Alcester, Stratford-upon-Avon and Warwick Junction Railway Bill, 21 March 1865.
21. Newspaper reports, *Northern Daily Express*, May–June 1861.

CHAPTER 5

1. Unless otherwise stated, the information on George Stephenson is taken from one of the following sources: Hunter Davies, *George Stephenson*, 1975; L.T.C. Rolt, *George and Robert Stephenson*, 1960; and Samuel Smiles, *The Lives of George and Robert Stephenson*, 1874 edition.
2. Sir John Rennie, *Autobiography*, 1875.
3. Olinthus J. Vignoles, *Life of Charles Blacker Vignoles*.
4. Nicholas Wood, *Treatise on Rail Roads*, 1825.
5. For a full account, see J.B. Snell, *Mechanical Engineering*: *Railways*, 1971.
6. Letter to William James, 20 December 1821.
7. Liverpool and Manchester Railway: Minutes of Evidence, 1825.
8. House of Lords Committee on the London and Birmingham Bill, June 1832.
9. London and Birmingham Railway: Minutes, 1835–8.
10. Edwin Clark, *The Britannia and Conway Tubular Bridges*, 1850.
11. Edwin Clark, *The Britannia and Conway Tubular Bridges*.
12. For two contrasting views of Brunel, see L.T.C. Rolt, *Isambard Kingdom Brunel*, 1951, and Adrian Vaughan, *Isambard Kingdom Brunel*, 1991.
13. Reports to the GWR, 1835–42; 15 September 1835.
14. Report to the GWR, 1 September 1836.
15. F.R. Conder, *Personal Recollections of English Engineers*, 1868 (reprinted 1983).
16. Letters from Stephenson to Brunel, 5 August 1838 and 25 February 1848.
17. Brunel's diary, 1832–40, entry for 5 August 1833.
18. Report to the GWR, 26 January 1837.

CHAPTER 6

1. See N.W. Webster, *Joseph Locke*, 1970.
2. London and Birmingham Railway, London Sub-Committee Minute Book, 12 December 1834.
3. See T. Mackay, *The Life of Sir John Fowler*.
4. East Lincolnshire Railway Papers, Agreement, 18 December 1846.
5. Agreement between East Lincolnshire Railway and John Waring, 1846.
6. Eden Valley Railway Papers, letter dated 15 February 1860.
7. William Cubitt's Report, 20 April 1848.
8. South Wales Railway Minute Book, November 1847.
9. I.K. Brunel, Private Letter Book, 16 January 1836.
10. Report to the Sheffield and Manchester Railway Company, 20 February 1837.
11. Olinthus J. Vignoles, *Life of Charles Blacker Vignoles*.
12. Report on the Bishop Auckland and Weardale Railway, 1841.
13. Frederick S. Williams, *Our Iron Roads*.
14. Report to the GWR, 19 November 1835.
15. Huddersfield and Manchester Railway and Canal Company Proceedings, 24 August 1846.
16. For a further account see Marcus Binney and David Pearce, *Railway Architecture*, 1979.
17. Evidence on the Great Western Railway Bill, House of Lords, June 1835.

CHAPTER 7

1. Peter Lecount, *A Practical Treatise on Railways*.
2. Richard Creed's letters (79 in all) are held in the PRO.
3. Bishop Auckland and Weardale Journal.
4. Daily journal of T. Pearson, Sheffield and Rotherham Railway, 1837–8.
5. F.R. Conder, *Personal Recollections of English Engineers*.
6. South Durham and Lancashire Union Railway Papers, letter from James B. Baldwin, 17 August 1858.
7. *The Bristol Mirror*, 14 January 1837.
8. I.K. Brunel, Private Letter Book, 15 June 1838.
9. Report to the GWR, 8 October 1835.

10. Eden Valley Railway Papers, letter from Thomas Bouch, 12 November 1857.
11. George Chetwynd to Trent Vale Railway, 27 December 1845.
12. Vale of Neath Railway Correspondence, letter to land valuer, 12 December 1848.
13. W.F. Fairbank letters, 16 February 1836.
14. W.F. Fairbank letters, October 1836.
15. Pamphlets on the Wharfedale Railway, 1861.
16. Letter to Bishop Auckland and Weardale Railway, 14 May 1840.
17. Letter to South Durham and Lancashire Union Railway, 1 May 1863.
18. Committee of Investigation into the Brandling Junction Railway, 1843.
19. Eden Valley Railway Letter Book, letters dated 20 October 1858 and 4 July 1860.

CHAPTER 8

1. Quoted in L.T.C. Rolt, *George and Robert Stephenson*, 1960.
2. 'Duties of Mr Walker, Resident Engineer', Leeds and Thirsk Railway Report Book, 30 May 1846.
3. 'Duties of Mr Walker, Resident Engineer', Leeds and Thirsk Railway Report Book, 2 June 1846.
4. Detailed information on this and other contracts can be found in Lawrence Popplewell, *A Gazetteer of the Railway Contractors and Engineers of Central England, 1830–1914*, 1986.
5. I.K. Brunel, Private Letter Book, 10 April 1837, on contracts for Taff Vale.
6. John Francis, *A History of the English Railway*.
7. Mansfield and Southwell Construction Committee Minute.
8. Rhondda and Swansea Bay Railway, Minute Book, 1882–6.
9. Darlington and Barnard Castle Railway, letter dated 12 April 1853.
10. Contract between the Newcastle and Berwick Railway and Benjamin Lawton and John Rush.
11. F.R. Conder, *Personal Recollections of English Engineers*.
12. Contract dated 1862.
13. Evidence to the Committee to Inquire into the Conditions of the Labourers Employed in the Construction of Railways, 1846 (hereafter referred to as 1846 Committee).
14. See Sir Arthur Helps, *Life and Labours of Thomas Brassey, 1805–1870*, 1888.

15. Letter from M. Ramsey to George Hudson, 28 June 1845.
16. Settle and Carlisle Construction Committee Minute Book, 30 September 1871.
17. Quoted in Popplewell, *A Gazetteer of the Railway Contractors and Engineers of Central England, 1830–1914.*
18. Chester and Holyhead Railway Papers, 1 April 1847.
19. *Felix Farley's Bristol Journal*, 19 September 1845.
20. Undated letter from Ralph Lawson to the Bishop Auckland and Weardale Railway.
21. Letter from Thomas Storey (engineer), 29 March 1824.
22. South Durham and Lancashire Union Letters, 1858–63.
23. I.K. Brunel, Private Letter Book, 10 April 1839.
24. The details come from Brunel's Reports to the GWR, 1835–42.
25. For a full account see David Brooke, 'The Great Commotion at Mickleton Tunnel', *JRCHS*, Vol. XXX, July 1990.
26. Taff Vale Railway Letters, 30 October 1837.
27. Letter dated 23 March 1838.

CHAPTER 9

1. Terry Coleman, *The Railway Navvies*, 1965.
2. Report from a town missionary visiting the Leeds and Thirsk Railway, November 1847.
3. London and Birmingham Railway: Contractors' Wages and Petty Cash Book, 1833–4.
4. The principal sources used were Pamela Horn, *Labouring Life in the Victorian Countryside*, 1976; G.E. Mingay, *Rural Life in Victorian England*, 1977; and Raphael Samuel, *Village Life and Labour*, 1975.
5. C. Henry Warren, *Happy Countryman*, 1939.
6. 1846 Committee.
7. Peter Lecount, *The History of the Railways Connecting London and Birmingham*, 1839.
8. *Illustrated London News*, 30 December 1854.
9. Police Inspector's Report, Leeds and Thirsk, 23 November 1847.
10. David Brooke, *The Railway Navvy*, 1983.
11. R.S. Joby, *The Railway Builders*, 1983.
12. Quoted in R.H.G. Thomas, *London's First Railway*, 1972.
13. Unless otherwise stated, the information comes from evidence given to the 1846 Committee.
14. Woodhead tunnel: Papers read before the Statistical Society of

Manchester, quoting a letter from John Robertson, 13 November 1845.

15. W.R. Mitchell, *The Railway Shanties*, 1975.
16. Settle and Carlisle Construction Committee Minute Book, 31 October 1871.
17. Leeds and Thirsk Railway Report Book, 22 February 1847.
18. Quoted in David Brooke, *The Railway Navvy*.

CHAPTER 10

1. Letter from Captain C.R. Moorsom, 26 May 1845.
2. *Felix Farley's Bristol Journal*, 28 April 1838.
3. *Preston Pilot*, 26 May 1838.
4. Taff Vale Railway, letter from George Bush, engineer, 27 September 1839.
5. Police Inspector's Report, 23 November 1847.
6. Unless otherwise stated, the following quotes and information are taken from evidence given to the 1846 Committee.
7. Woodhead tunnel: Papers read before the Statistical Society of Manchester.
8. Thomas Nicholson, 'Strictures on a Pamphlet', 1846.
9. Trent Valley Railway Letters, 1845–6.
10. *Felix Farley's Bristol Journal*, 7 February 1846.
11. *Felix Farley's Bristol Journal*, 16 May 1846.
12. John Francis, *A History of the English Railway*.
13. The Vicar of Batley to the Leeds and Dewsbury Railway Company, 1 May 1846.
14. *Felix Farley's Bristol Journal*, 13 July 1839.
15. Chester and Holyhead Railway, Resident Engineer's Letters, 13 February 1845.

CHAPTER 11

1. Charles Dickens, *Dombey and Son*, 1846.
2. J.C. Bourne, *Drawings of the London and Birmingham Railway*, 1839.
3. Frederick S. Williams, *Our Iron Roads*.
4. Robert Rawlinson, engineer to the Bridgewater Trust, Woodhead Papers.

5. Sir Arthur Helps, *Life and Labours of Thomas Brassey, 1805–1870.*
6. Frederick S. Williams, *Our Iron Roads.*
7. 1846 Committee.
8. *Felix Farley's Bristol Journal*, 8 August 1840.
9. Leeds and Thirsk Railway Report Book, 26 November 1846.
10. Part of a full account of a visit to Box tunnel in *Felix Farley's Bristol Journal*, 27 July 1839.
11. 1846 Committee.
12. 1846 Committee.
13. Thomas A. Walker, *The Severn Tunnel*, 1888.
14. Official Report, 17 July 1835, and *Annual Register*, 1835.
15. *Felix Farley's Bristol Journal*, 30 June 1838, 18 August 1838, 24 November 1838 and 9 February 1839.
16. 1846 Committee.
17. Leeds and Thirsk Railway Report Book, December 1846.
18. 1846 Committee.
19. David Brooke, 'The Other Cost of Railway Building', *JRCHS*, Vol. 29, July 1989.
20. Chester and Holyhead Railway Letters, 4 October 1846.
21. Robert Stephenson, Report, 6 April 1836, quoted in G.Y. Hemingway, 'Kilsby Tunnel', *JRCHS*, Vol. 26, March 1980.
22. Standing Orders, Exeter, Plymouth and Devonport Railway, 1837.
23. *Annual Review*, 1848.
24. Colin Maggs and Peter Nicholson, *The Honeybourne Line*, 1985.
25. Anthony Murray, *The Forth Railway Bridge*, 1983.
26. Middlesborough and Redcar Railway Paper, 2 December 1845.

CHAPTER 12

1. John Scott-Morgan, *The Colonel Stephens Railways*, 1978.
2. For a fuller account see Anthony Burton and John Scott-Morgan, *Britain's Light Railways*, 1985.
3. See L.T. Catchpole, *The Lynton & Barnstaple Railway*, 1936.
4. *Felix Farley's Bristol Journal*, 27 August 1836.
5. Fanny Kemble, *Records of a Girlhood*, 1878.
6. Accounts of both openings are given in Anthony Burton, *The Rainhill Story*, 1980.

7. Trent Valley Railway Letters, opening ceremony at Atherstone, 22 November 1846.
8. Edward Churton, *The Rail Road Book of England*, 1851.
9. George Measom, *The Illustrated Guide to the Great Western Railway*, 1852.

Index